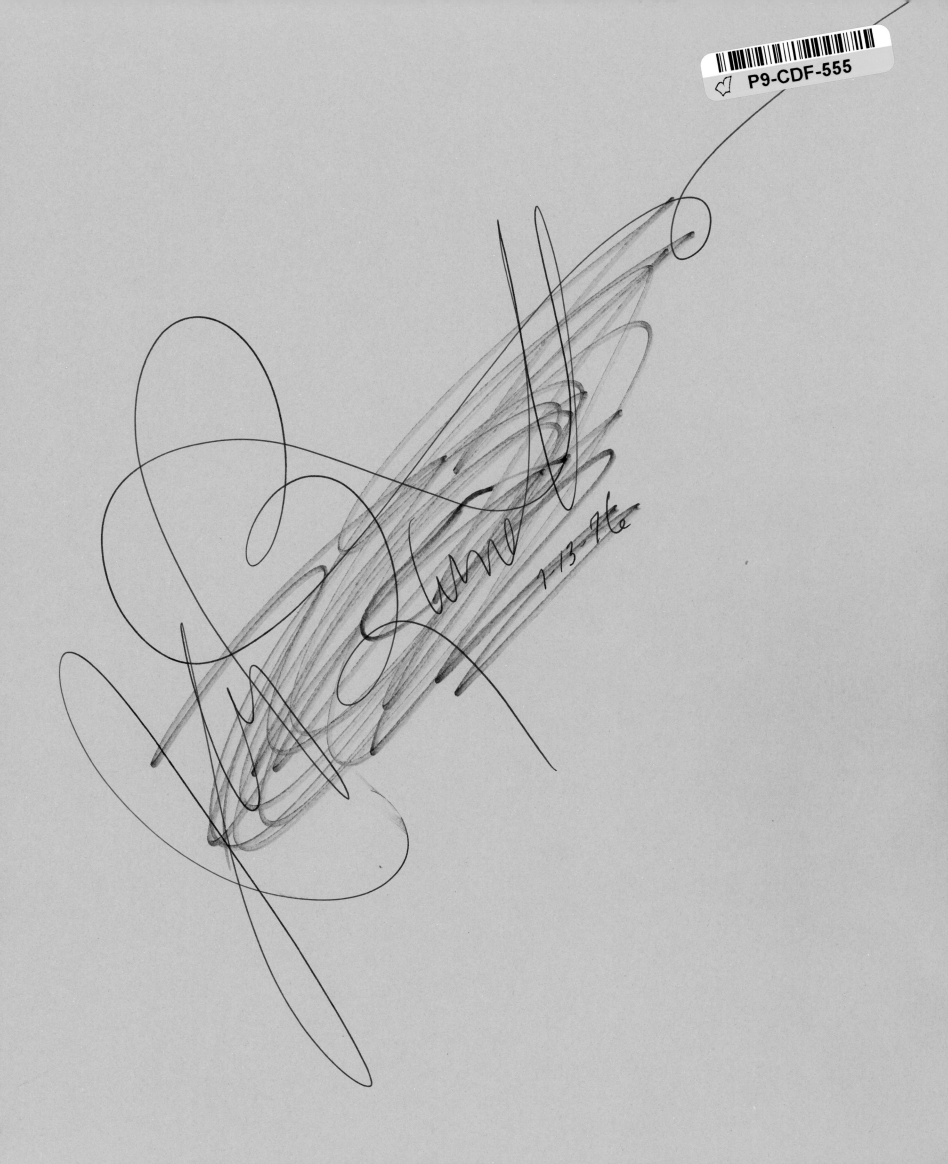

JOHN MUIR'S LONGEST WALK

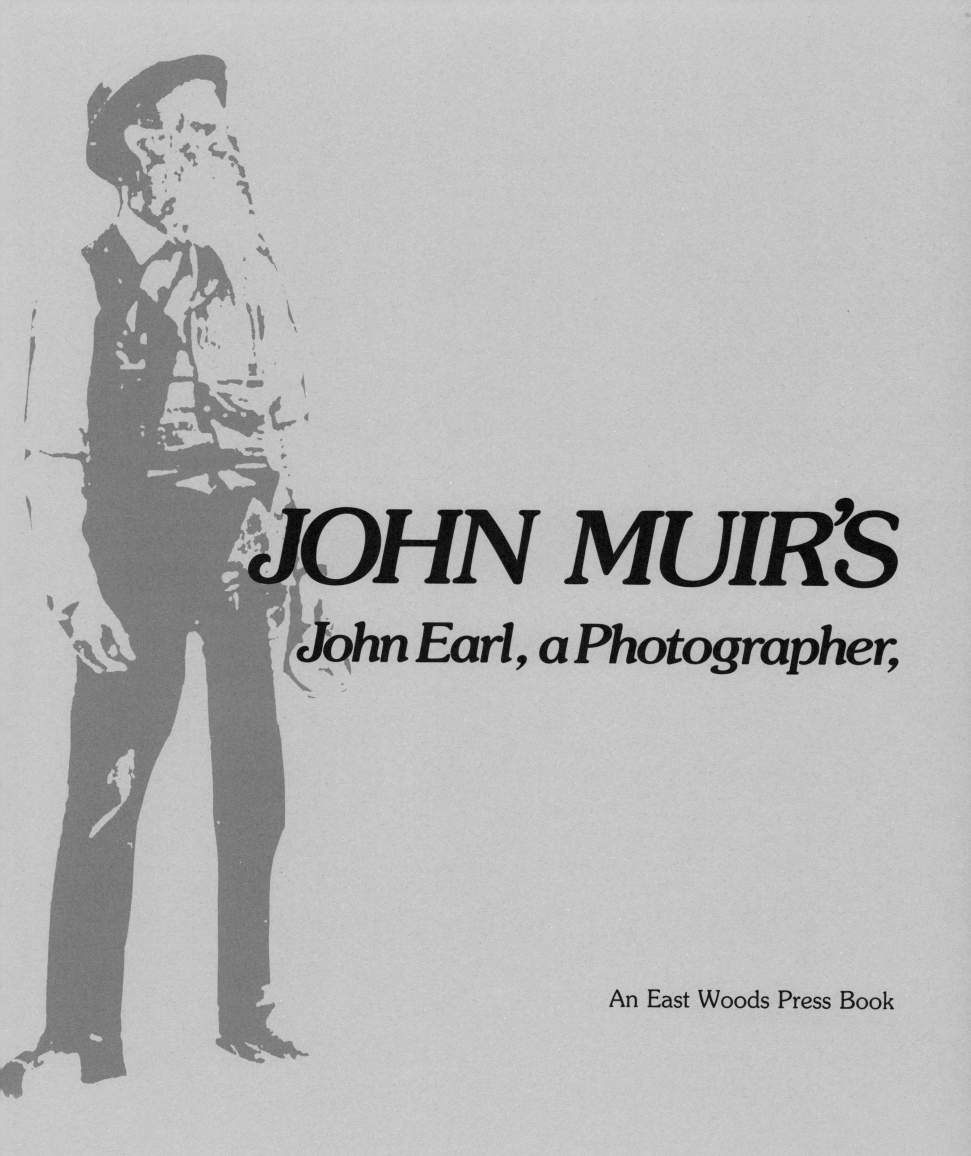

JOHN MUIR'S
John Earl, a Photographer,

An East Woods Press Book

LONGEST WALK

Traces His Journey to Florida

With Excerpts from John Muir's *Thousand-Mile Walk to the Gulf*

Doubleday & Company, Inc.
Garden City, New York
1975

International Standard Book Number: 0-385-09216-4   Library of Congress Catalog Card Number: 75-14855
First Edition in the United States of America

Conceived and Produced by East Woods Press, Inc., New York
Printed and bound in Italy by Mondadori Editore, Verona
Design by Scott Chelius   Cartography by Andrew Mudryk   Typography by Harold Black Inc.

The original work by John Muir, *A Thousand-Mile Walk to the Gulf*, from which the contents of this edition are
excerpted, was published in 1916 by Houghton Mifflin Company. Reprinted by permission of Houghton Mifflin Company.

To my wife, Leonora,
without whose love, patience, and understanding
this book could not have been possible.

About this book

At first John Earl was merely a name on a scrap of paper in a "Photographers" file when I was gathering pictures for a wilderness calendar. Soon he was a voice on the telephone—pleasant and musical—describing the beauty of North Carolina's Linville Gorge. Then he became a courteous visitor to New York City, eager to speculate about forthcoming adventures tracing the path taken by John Muir in 1867 on his epic thousand-mile walk from Kentucky to Florida. . . . Now, several years later, John Earl is a good friend of every person at East Woods Press, and we are proud to have the privilege of presenting his thousand-mile photographs to the public.

A few sentences of explanation: The main text of *John Muir's Longest Walk* is by Muir himself. It was excerpted by us from his *Thousand-Mile Walk to the Gulf*, which was published in 1916, two years after his death. The introduction, also from the 1916 volume, is by William Frederic Badè, who edited the original book making use of Muir's journals and various other research materials in a highly intelligent way. The preface, comments on Muir's route, picture captions, and, of course, magnificent photographs are by John Earl. For obvious reasons we have omitted the final three chapters of *A Thousand-Mile Walk*: Muir's sojourn in Cuba, his "crooked" route to the West Coast, via New York City, and his first year in California. Muir aficionados and everybody else curious about those chapters and about what we have eliminated with all the ellipses are urged to acquire a copy of the original book; it has been available since 1969 thanks to Norman S. Berg, publisher, of Dunwoody, Georgia.

We hope that *John Muir's Longest Walk* will bring readers as much joy as we have had in putting it together, as John Earl had in his labor of love and, indeed, as John Muir had in those extraordinary months in the wilderness in 1867.

Constance Stallings, Editor
East Woods Press
April, 1975

Contents

Preface

I have known about John Muir for most of my life. I knew he was the father of conservation, the founder of the Sierra Club and the explorer of Yosemite, Sequoia, Kings Canyon, Mount Rainier, the Petrified Forest, the Grand Canyon and Glacier Bay. But what I did not know until 1969 when I became involved in nature photography was that Muir's first long walk of any consequence took place in the East. In 1867 at the age of twenty-nine he walked for a distance of one thousand miles, from Louisville, Kentucky, to Cedar Key, Florida, in less than two months. I learned that he had kept a journal of the walk and that it had been published in the form of a book entitled *A Thousand-Mile Walk to the Gulf*. I was fortunate enough to obtain a reprint of the original and became fascinated by what I read.

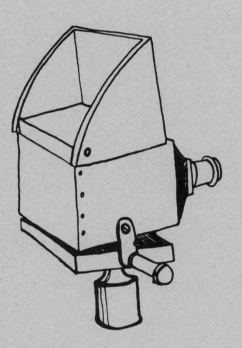

In 1972 my friend David Brower told me that someone in Kentucky had thought it would be wonderful for a nature photographer to retrace Muir's thousand-mile walk, taking pictures for a book depicting the things Muir had seen along the way. Brower said he believed that I was the perfect choice for the project.

Since I already had great admiration for John Muir, I slowly began to formulate in my mind the trip which soon became a reality—an experience which has changed my life and given me a far truer understanding of Muir, of the South, where I was born and grew up, and of myself.

I began photographing the thousand-mile walk in March of 1973,

13

starting at Cedar Key where Muir ended his walk and retracing his route backward toward Kentucky so as to follow spring north.

I located the site where Muir worked for a few days in a sawmill before catching malaria, and the place where he lived while recuperating. I sailed among the Keys and found many of the things he wrote about: "long-winged gulls," "burnished, treeless plain" at the Gulf and "many gems of palmy islets." I "sat beneath a moss-draped live-oak watching birds feeding on the shore," saw palmettos, ferns, mosses growing on live-oaks, saw mangroves and rushes of all types, and watched pelicans in great numbers. But with great sorrow I also saw something else—the effect man has had on Florida during the 106 years since John Muir arrived there. And I realized that it was going to be extremely difficult to find photographs depicting the unspoiled natural beauty that existed in the East in 1867.

One day I suddenly realized that the only way I could succeed would be to think of myself as being John Muir, in the body of John Earl, and to learn to see things through my eyes in the same way he saw them through his. Not long before Muir started the thousand-mile walk, he had suffered an eye injury which left him temporarily blinded. As a result of that accident, he promised himself and God that if his sight returned he would devote his life to God's work of studying and protecting nature. This he did indeed.

From then on, my work became easier. I pretended that I was John Muir, searching for nature's secrets, seeing this thing and that for the first time and shutting out the unnatural. I began to realize that I must occasionally deviate from the original trail to avoid shopping centers, subdivisions, trailer parks, factories and all the other developments that man has labeled "progress." Thanks to Muir, a few places do remain almost as he first found them. They were the pockets I sought out—places such as Florida's Cedar Keys National Wildlife Refuge, San Felasco Hammock, Lake Orange, the Santa Fe and Ichetucknee rivers, quiet inland woods and remote Gulf marshes.

As I moved north into Georgia, I photographed Ossabaw Island with its unspoiled beaches, vast marshes, virgin forests and wealth of wild flora and fauna. Then I visited the bluffs of the Savannah River and, farther north, found rivers and mountains still in their natural state.

As I worked my way into North Carolina, I began to see things more and more through the eyes of Muir. And I began to understand what he meant when he wrote about his "eagerness . . to baptize all of my fellow sinners in the beauty of God's mountains." I learned to "consider the lilies how they grow," a Bible passage Muir once used to explain to a mountain man the importance of what he called "doing God's work."

One day in North Carolina I learned what Muir meant when he described the experience of being caught in a violent thunderstorm as one of sheer ecstasy and beauty. Thunder, lightning, rain and hail engulfed me, and the trees bent over almost to the ground under the force of the strongest wind I have ever experienced. At first, as John Earl, I felt disappointed at having to wrap my cameras in plastic bags and run for shelter. Then, as the storm became more fierce, I began to fear for my safety. Then I suddenly "became" John Muir, and my only emotion was a desire to be a part of the great thing which was unfolding around me, to allow myself to be caught up in the divine exhilaration of nature. It was one of the most profound experiences of my life. When the storm ended, I discovered that I had walked for three miles with sixty pounds of equipment on my back, and gained 1,500 feet in elevation, without once stopping to rest or even feeling tired. Wet to the skin, I felt only a rare contentment. From that day forward the project became not a task but a supreme pleasure.

Entering Tennessee and Kentucky, I grew increasingly excited. I knew that Muir had visited Mammoth Cave, so I, too, visited Mammoth Cave. Going north toward Louisville, I walked through dense green forests and past knobby hills where rock formations reached up like towers to the sky . . . All summer I returned again and again to places so beautiful it seemed impossible that I could ever photograph them all.

By September when it was again time for me to follow the footsteps of John Muir, I was more than ready. On this trip I began in Kentucky so as to follow fall southward. I traveled from day to day filled with the delight of knowing that both Muir and I had been there before. I discovered new places, too, ones I never knew existed. Sometimes I was right into Muir's footsteps. Once I cracked a rib leaning over a cliff to photograph the changing light in a valley below. Another time I stubbed my toe so badly it grew to twice its size. In North Carolina, running to get more film to capture a sunset, I fractured my skull. But these accidents seemed only minor inconveniences compared with the delight of discovering nature's mysteries.

By late November when I reached the Gulf, and had traveled more than five thousand miles, I was a different John Earl. I had learned to see both nature and myself in a new way, through the eyes of a man I will never know, who taught me not only how to see but also how to feel. John Muir died in 1914 at the age of seventy-six, but he lives on today within me. I shall be forever indebted to him.

There are other debts. Muir did much for all of us. He helped guide the Antiquities Act of 1906, which gave presidents the power to establish national monuments. He worked for the creation of the National Park Service; it became a reality two years after his death. He influenced Teddy Roosevelt to begin setting aside land; that influence eventually resulted in five new national parks, sixteen national monuments and 148 million acres of national forest. Every person in America owes John Muir a debt which can be paid only by our joint efforts to protect, maintain and preserve the small amount of precious wilderness that still remains.

I sincerely hope that this book will serve not only as a beautiful record of John Muir's thousand-mile walk but also as proof to every reader that he is a citizen of the universe. May we all pause for a moment and try to understand what Muir meant when he said: "Oh, these vast, calm, measureless mountain days . . . in whose light everything seems equally divine, opening a thousand windows to show us God." Perhaps then we will better understand what he felt when he wrote: "I set forth . . . joyful and free, on a thousand-mile walk to the Gulf of Mexico . . . . by the wildest, leafiest, and least trodden way I could find."

John Earl
Chamblee, Georgia
March, 1975

Introduction

"John Muir, Earth-planet, Universe."—These words are written on the inside cover of the notebook from which the contents of this volume have been taken. They reflect the mood in which the late author and explorer undertook his thousand-mile walk to the Gulf of Mexico a half-century ago. No less does this refreshingly cosmopolitan address . . . reveal the temper and the comprehensiveness of Mr. Muir's mind. He never was and never could be a parochial student of nature. Even at the early age of twenty-nine his eager interest in every aspect of the natural world had made him a citizen of the universe.

While this was by far the longest botanical excursion which Mr. Muir made in his earlier years, it was by no means the only one. He had botanized around the Great Lakes, in Ontario, and through parts of Wisconsin, Indiana, and Illinois. On these expeditions he had disciplined himself to endure hardship, for his notebooks disclose the fact that he often went hungry and slept in the woods, or on the open prairies, with no cover except the clothes he wore.

"Oftentimes," he writes in some unpublished biographical notes, "I had to sleep out without blankets, and also without supper or breakfast. But usually I had no great difficulty in finding a loaf of bread in the widely scattered clearings of the farmers. With one of these big backwoods loaves I was able to wander many a long, wild mile, free as the winds in the glorious forests and bogs, gathering plants and feeding on God's abounding, inexhaustible spiritual beauty bread. Only once in my long Canada wanderings was the deep peace of the wilderness savagely broken. It happened in the maple woods about midnight, when I was cold and my fire was low. I was awakened by the awfully dismal howling of the wolves, and got up in haste to replenish the fire."

It was not, therefore, a new species of adventure upon which Mr. Muir embarked when he started on his Southern foot-tour. It was only a new response to the lure of those favorite studies which he had already pursued over uncounted miles of virgin Western forests and prairies. Indeed, had it not been for the accidental injury to his right eye in the month of March, 1867, he probably would have started somewhat earlier than he did. In a letter written to Indianapolis friends on the day after the accident, he refers mournfully to the interruption of a long-cherished plan. "For weeks," he writes, "I have daily consulted maps in locating a route through the Southern States, the West Indies, South America, and Europe — a botanical journey studied for years. And so my mind has long been in a glow with visions of the glories of a tropical flora; but, alas, I am half blind. My right eye, trained to minute analysis, is lost and I have scarce heart to open the other. Had this journey been accomplished, the stock of varied beauty acquired would have made me willing to shrink into any corner of the world, however obscure and however remote."

The injury to his eye proved to be less serious than he had at first supposed. In June he was writing to a friend: "I have been reading and botanizing for some weeks, and find that for such work I am not very much disabled. I leave this city [Indianapolis] for home to-morrow, accompanied by Merrill Moores, a little friend of mine. We will go to Decatur, Illinois, thence northward through the wide prairies, botanizing a few weeks by the way. . . . I hope to go South towards the end of the summer, and as this will be a journey that I know very little about, I hope to profit by your counsel before setting out."

In an account written after the excursion he says: "I was eager to see Illinois prairies on my way home, so we went to Decatur, near the center of the State, thence north [to Portage] by Rockford and Janesville. I botanized one week on the prairie about seven miles southwest of Pecatonica. . . . To me all plants are more precious than before. My poor eye is not better, nor worse. A cloud is over it, but in gazing over the widest landscapes, I am not always sensible of its presence."

By the end of August Mr. Muir was back again in Indianapolis. He had found it convenient to spend a "botanical week" among his University friends in Madison. So keen was his interest in plants at this time that an interval of five hours spent in Chicago was promptly turned to account in a search for them. "I did not find many plants in her tumultuous streets," he complains; "only a few grassy plants of wheat, and two or three species of weeds, — amaranth, purslane, carpet-weed, etc., — the weeds, I suppose, for man to walk upon, the wheat to feed him. I saw some green algae, but no mosses. Some of the latter I expected to see on wet walls, and in seams on the pavements. But I suppose that the manufacturers' smoke and the terrible noise are too great for the hardiest of them. I wish I knew where I was going. Doomed to be 'carried of the spirit into the wilderness,' I suppose. I wish I could be more moderate in my desires, but I cannot, and so there is no rest."

The letter noted above was written only two days before he started on his long walk to Florida. If the concluding sentences still reflect indecision, they also convey a hint of the overmastering impulse under which he was acting. The opening sentences of his journal, afterwards crossed out, witness to this sense of inward compulsion which he felt. "Few bodies," he wrote, "are inhabited by so satisfied a soul that they are allowed exemption from extraordinary exertion through a whole life. . . . For many a year I have been impelled toward the Lord's tropic gardens of the South. Many influences have tended to blunt or bury this constant longing, but it has outlived and overpowered them all."

Muir's love of nature was so largely a part of his religion that he naturally chose Biblical phraseology when he sought a vehicle for his feelings. No prophet of old could have taken his call more seriously, or have entered upon his mission more fervently. During the long days of his confinement in a dark room he had opportunity for much reflection. He concluded that life was too brief and uncertain, and time too precious, to waste upon belts and saws; that while he was pottering in a wagon factory, God was making a world; and he determined that, if his eyesight was spared, he would devote the remainder of his life to a study of the process. Thus the previous bent of his habits and studies, and the sobering thoughts induced by one of the bitterest experiences of his life, combined to send him on the long journey recorded in these pages.

Some autobiographical notes found among his papers furnish interesting additional details about the period between his release from the dark room and his departure for the South. "As soon as I got out into heaven's light," he says, "I started on another long excursion, making haste with all my heart to store my mind with the Lord's beauty, and thus be ready for any fate, light or dark. And it was from this time that my long, continuous wanderings may be said to have fairly commenced. I bade adieu to mechanical inventions, determined to devote the rest of my life to the study of the inventions of God. I first went home to Wisconsin, botanizing by the way, to take leave of my father and mother, brothers and sisters, all of whom were still living near Portage. I also visited the neighbors I had known as a boy, renewed my acquaintance with them after an absence of several years, and bade each a formal good-bye. When they asked where I was going I said, 'Oh! I don't know — just anywhere in the wilderness, south-ward. I have already had glorious glimpses of the Wisconsin, Iowa, Michigan, Indiana, and Canada wildernesses; now I propose to go South and see something of the vegetation of the warm end of the country, and if possible to wander far enough into South America to see tropical vegeta-tion in all its palmy glory.'

"The neighbors wished me well, advised me to be careful of my health, and reminded me that the swamps in the South were full of malaria. . . ."

The formal leave-taking from family and neighbors indicates his belief that he was parting from home and friends for a long time. On Sunday, the 1st of September, 1867, Mr. Muir said good-bye also to his Indianapolis friends, and went by rail to Jeffersonville, where he spent the night. The next morning he crossed the river, walked through Louisville, and struck southward through the State of Kentucky. A letter written a week later "among the hills of Bear Creek, seven miles southeast of Burkesville, Kentucky," shows that he had covered about twenty-five miles a day. "I walked from Louisville," he says, "a distance of one hundred and seventy miles, and my feet are sore. But, oh! I am paid for all my toil a thousand times over. I am in the woods on a hilltop with my back against a moss-clad log. I wish you could see my last evening's bedroom. The sun has been among the tree-tops for more than an hour; the dew is nearly all taken back, and the shade in these hill basins is creeping away into the unbroken strongholds of the grand old forests.

"I have enjoyed the trees and scenery of Kentucky exceedingly. How shall I ever tell of the miles and miles of beauty that have been flowing into me in such measure? These lofty curving ranks of lobing, swelling hills, these concealed valleys of fathomless verdure, and these lordly trees with the nursing sunlight glancing in their leaves upon the outlines of the magnificent masses of shade embosomed among their wide branches — these are cut into my memory to go with me forever.

"I was a few miles south of Louisville when I planned my journey. I spread out my map under a tree and made up my mind to go through Kentucky, Tennessee, and Georgia to Florida, thence to Cuba, thence to some part of South America; but it will be only a hasty walk. I am thankful, however, for so much. My route will be through Kingston and Madisonville, Tennessee, and through Blairsville and Gainesville, Georgia. Please write me at Gainesville. I am terribly letter-hungry. I hardly dare to think of home and friends."

In editing the journal I have endeavored, by use of all the available evidence, to trail Mr. Muir as closely as possible on maps of the sixties as well as on the most recent state and topographical maps. The one used by him has not been found, and probably is no longer in existence. Only about twenty-two towns and cities are mentioned in his journal. This constitutes a very small number when one considers the distance he covered. Evidently he was so absorbed in the plant life of the region traversed that he paid no heed to towns, and perhaps avoided them wherever possible.

The sickness which overtook him in Florida was probably of a malarial kind, although he describes it under different names. It was, no doubt, a misfortune in itself, and a severe test for his vigorous constitution. But it was also a blessing in disguise, inasmuch as it prevented him from carrying out his foolhardy plan of penetrating the tropical jungles of South America along the Andes to a tributary of the Amazon, and then floating down the river on a raft to the Atlantic. As readers of the journal will perceive, he clung to this intention even during his convalescence at Cedar Keys and in Cuba. In a letter dated the 8th of November he describes himself as "just creeping about getting plants and strength after my fever." Then he asks his correspondent to direct letters to New Orleans, Louisiana. "I shall have to go there," he writes, "for a boat to South America. I do not yet know to which point in South America I had better go." His hope to find there a boat for South America explains an otherwise mystifying letter in which he requested his brother David to send him a certain sum of money by American Express order to New Orleans. As a matter of fact he did not go into Louisiana at all, either because he learned that no south-bound ship was available at the mouth of the Mississippi, or because the unexpected appearance of the Island Belle in the harbor of Cedar Keys caused him to change his plans.

In later years Mr. Muir himself strongly disparaged the wisdom of his plans with respect to South America. . . . Nevertheless the Andes and the South American forests continued to fascinate his imagination, as his letters

21

show, for many years after he came to California. When the long deferred journey to South America was finally made in 1911, forty-four years after the first attempt, he whimsically spoke of it as the fulfillment of those youthful dreams that moved him to undertake his thousand-mile walk to the Gulf.

Mr. Muir always recalled with gratitude the Florida friends who nursed him through his long and serious illness. In 1898, while traveling through the South on a forest-inspection tour with his friend Charles Sprague Sargent, he took occasion to revisit the scenes of his early adventures. It may be of interest to quote some sentences from letters written at that time to his wife and to his sister Sarah. "I have been down the east side of the Florida peninsula along the Indian River," he writes, "through the palm and pine forests to Miami, and thence to Key West and the southmost keys stretching out towards Cuba. Returning, I crossed over to the west coast by Palatka to Cedar Keys, on my old track made thirty-one years ago, in search of the Hodgsons who nursed me through my long attack of fever. Mr. Hodgson died long ago, also the eldest son, with whom I used to go boating among the keys while slowly convalescing."

He then tells how he found Mrs. Hodgson and the rest of the family at Archer. They had long thought him dead and were naturally very much surprised to see him. Mrs. Hodgson was in her garden and he recognized her, though the years had altered her appearance. Let us give his own account of the meeting: "I asked her if she knew me. 'No, I don't,' she said; 'tell me your name.' 'Muir,' I replied. 'John Muir? My California John Muir?' she almost screamed. I said, 'Yes, John Muir; and you know I promised to return and visit you in about twenty-five years, and though I am a little late — six or seven years — I've done the best I could.' The eldest boy and girl remembered the stories I told them, and when they read about the Muir Glacier they felt sure it must have been named for me. I stopped at Archer about four hours, and the way we talked over old times you may imagine." From Savannah, on the same trip, he wrote: "Here is where I spent a hungry, weary, yet happy week camping in Bonaventure graveyard thirty-one years ago. Many changes, I am told, have been made in its graves and avenues of late, and how many in my life!"

In perusing this journal the reader will miss the literary finish which Mr. Muir was accustomed to give to his later writings. This fact calls for no excuse. Not only are we dealing here with the earliest product of his pen, but with impressions and observations written down hastily during pauses in his long march. He apparently intended to use this raw material at some time for another book. If the record, as it stands, lacks finish and adornment, it also possesses the immediacy and the freshness of first impressions.

The sources which I have used in preparing this volume are threefold: (1) the original journal, of which the first half contained many interlinear revisions and expansions. . . . (2) a wide-spaced, typewritten, rough copy of the journal, apparently in large part dictated to a stenographer; it is only slightly revised, and comparison with the original journal shows many significant omissions and additions; (3) two separate elaborations of his experiences in Savannah when he camped there for a week in the Bonaventure graveyard. Throughout my work upon the primary and secondary materials I was impressed with the scrupulous fidelity with which he adhered to the facts and impressions set down in the original journal.

Readers of Muir's writings need scarcely be told that this book, autobiographically, bridges the period between *The Story of my Boyhood and Youth* and *My First Summer in the Sierra*. . . .

William Frederic Badè

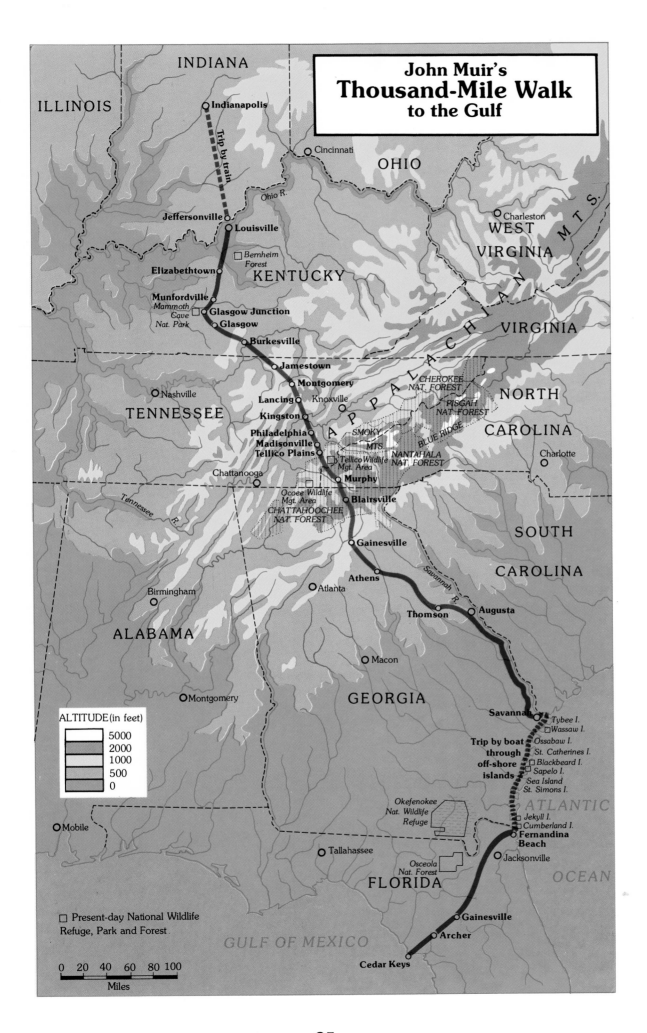

ILLINOIS

INDIANA

○ Indianapolis

Trip by train

○ Cincinnati

OHIO

Ohio R.

○ Charleston

WEST
VIRGINIA

Jeffersonville ○—○ Louisville

□ *Bernheim
Forest*

Elizabethtown ○

KENTUCKY

VIRGINIA

Munfordville ○
*Mammoth
Cave
Nat. Park* □ ○ Glasgow Junction
Glasgow ○
Burkesville ○

○ Jamestown

○ Montgomery

Lancing ○   Knoxville
○ Nashville   ○

TENNESSEE

Kingston ○

Philadelphia ○
Madisonville ○
Tellico Plains ○

A P P A L A C H I A N   M T S.

CHEROKEE
NAT. FOREST

PISGAH
NAT FOREST

NORTH
CAROLINA

SMOKY
MTS.   BLUE RIDGE

□ *Tellico Wildlife
Mgt. Area*   NANTAHALA
NAT. FOREST

Chattanooga ○   *Ocoee Wildlife
Mgt. Area*   ○ Murphy

○ Charlotte

CHATTAHOOCHEE
NAT. FOREST   ○ Blairsville

○ Gainesville

SOUTH
CAROLINA

Tennessee R.

○ Birmingham   ○ Atlanta

Athens ○

Savannah R.

Thomson ○   ○ Augusta

ALABAMA

○ Macon

GEORGIA

○ Montgomery

ALTITUDE (in feet)

5000
2000
1000
500
0

○ Mobile

*Okefenokee
Nat. Wildlife
Refuge*

Savannah ○   *Tybee I.*
□ *Wassaw I.*

*Trip by boat
through
off-shore
islands*   *Ossabaw I.*
St. Catherines I.
□ *Blackbeard I.*
□ *Sapelo I.*
*Sea Island
St. Simons I.*

ATLANTIC

□ *Jekyll I.*
□ *Cumberland I.*
○ Fernandina
Beach

○ Tallahassee

*Osceola
Nat. Forest*

○ Jacksonville

OCEAN

FLORIDA

□ Present-day National Wildlife
Refuge, Park and Forest

○ Gainesville
○ Archer

GULF OF MEXICO

○ Cedar Keys

0 20 40 60 80 100

Miles

John Muir's
Thousand-Mile Walk
to the Gulf

1
Kentucky Forests and Caves

I had long been looking from the wild woods and gardens of the Northern States to those of the warm South, and at last, all drawbacks overcome, I set forth [from Indianapolis] on the first day of September, 1867, joyful and free, on a thousand-mile walk to the Gulf of Mexico. [The trip to Jeffersonville, on the banks of the Ohio, was made by rail.] Crossing the Ohio at Louisville [September 2], I steered through the big city by compass without speaking a word to any one. Beyond the city I found a road running southward, and after passing a scatterment of suburban cabins and cottages I reached the green woods and spread out my pocket map to rough-hew a plan for my journey.

My plan was simply to push on in a general southward direction by the wildest, leafiest, and least trodden way I could find, promising the greatest extent of virgin forest. Folding my map, I shouldered my little bag and plant press and strode away among the old Kentucky oaks, rejoicing in splendid visions of pines and palms and tropic flowers in glorious array, not, however, without a few cold shadows of loneliness, although the great oaks seemed to spread their arms in welcome.

When John Muir left Louisville, he walked south toward Mammoth Cave. Today his route is spanned almost exactly by Interstate Highway 65. Many of the fine oaks have long since been cut; much of the land has recently given way to subdivisions, shopping centers, and parking lots. So to photograph places that look like the country Muir saw, I had to deviate from his path.

I have seen oaks of many species in many kinds of exposure and soil, but those of Kentucky excel in grandeur all I had ever before beheld. They are broad and dense and bright green. In the leafy bowers and caves of their long branches swell magnificent avenues of shade, and every tree seems to be blessed with a double portion of strong exulting life. Walked twenty miles, mostly on river bottom, and found shelter in a rickety tavern.

29

September 3. Escaped from the dust and squalor of my garret bedroom to the glorious forest. All the streams that I tasted hereabouts are salty and so are the wells. Salt River was nearly dry. Much of my way this forenoon was over naked limestone. After passing the level ground that extended twenty-five or thirty miles from the river I came to a region of rolling hills called Kentucky Knobs — hills of denudation, covered with trees to the top. Some of them have a few pines. For a few hours I followed the farmers' paths, but soon wandered away from roads and encountered many a tribe of twisted vines difficult to pass. . . .

In my search for wildness I discovered many ancient Indian carvings like these in the Red River Gorge. It saddens me to know that they and so much else of beauty may soon be lost forever, drowned under the reservoir of a dam the U.S. Army Corps of Engineers intends to build on the Red River.

September 4. The sun was gilding the hilltops when I was awakened by the alarm notes of birds whose dwelling in a hazel thicket I had disturbed. They flitted excitedly close to my head, as if scolding or asking angry questions, while several beautiful plants, strangers to me, were looking me full in the face. The first botanical discovery in bed! This was one of the most delightful camp grounds, though groped for in the dark, and I lingered about it enjoying its trees and soft lights and music.

This wooded trail at Yahoo Falls in Daniel Boone National Forest is typical of many trails that Muir followed on his walk through Kentucky. Here are the towering oaks and leafy, green hardwoods he so admired.

Walked ten miles of forest. Met a strange oak with willow-looking leaves. Entered a sandy stretch of black oak called ''Barrens,'' many of which were sixty or seventy feet in height, and are said to have grown since the fires were kept off, forty years ago.

September 5. No bird or flower or friendly tree above me this morning; only squalid garret rubbish and dust. Escaped to the woods. Came to the region of caves. At the mouth of the first I discovered, I was surprised to find ferns which belonged to the coolest nooks of Wisconsin and northward, but soon observed that each cave rim has a zone of climate peculiar to itself, and it is always cool. This cave had an opening about ten feet in diameter, and twenty-five feet perpendicular depth. A strong cold wind issued from it and I could hear the sounds of running water. A long pole was set against its walls as if intended for a ladder, but in some places it was slippery and smooth as a mast and would test the climbing powers of a monkey. The walls and rim of this natural reservoir were finely carved and flowered. Bushes leaned over it with shading leaves, and beautiful ferns and mosses were in rows and sheets on its slopes and shelves. Lingered here a long happy while, pressing specimens and printing this beauty into memory. . . .

September 6. Started at the earliest bird song in hopes of seeing the great Mammoth Cave before evening. . . .

Arrived at Horse Cave, about ten miles from the great cave. The entrance is by a long easy slope of several hundred yards. It seems like a noble gateway to the birthplace of springs and fountains and the dark treasuries of the mineral kingdom. This cave is in a village [of the same name] which it supplies with an abundance of cold water, and cold air that issues from its fern-clad lips. In hot weather crowds of people sit about it in the shade of the trees that guard it. This magnificent fan is capable of cooling everybody in the town at once.

Those who live near lofty mountains may climb to cool weather in a day or two, but the overheated Kentuckians can find a patch of cool climate in almost every glen in the State. The villager who accompanied me said that Horse Cave had never been fully explored, but that it was several miles in length at least. He told me that he had never been at Mammoth Cave — that it was not worth going ten miles to see, as it was nothing but a hole in the ground, and I found that his was no rare case. He was one of the useful, practical men — too wise to waste precious time with weeds, caves, fossils, or anything else that he could not eat.

Arrived at the great Mammoth Cave. I was surprised to find it in so complete naturalness. A large hotel with fine walks and gardens is near it. But fortunately the cave has been unimproved, and were it not for the narrow trail that leads down the glen to its door, one would not know that it had been visited. There are house-rooms and halls whose entrances give but slight hint of their grandeur. And so also this magnificent hall in the mineral kingdom of Kentucky has a door comparatively small and unpromising. One might pass within a few yards of it without noticing it. A strong cool breeze issues constantly from it, creating a northern climate for the ferns that adorn its rocky front.

I never before saw Nature's grandeur in so abrupt contrast with paltry artificial gardens. The fashionable hotel grounds are in exact parlor taste,

This trail at Natural Bridge State Park was so narrow I had to walk sideways to get through with my backpack on.

A sandstone wall near Gray's Arch—a small part of one of the finest formations in the Red River Gorge.

with many a beautiful plant cultivated to deformity, and arranged in strict geometrical beds, the whole pretty affair a laborious failure side by side with Divine beauty. The trees around the mouth of the cave are smooth and tall and bent forward at the bottom, then straight upwards. Only a butternut seems, by its angular knotty branches, to sympathize with and belong to the cave, with a fine growth of *Cystopteris* and *Hypnum*.

Nowhere in Kentucky did I feel more at home than at Natural Arch in Daniel Boone National Forest, which is said to have been a hunting ground for Shawnee Indians. This great arch of sandstone is one of many in the Cumberland Mountains.

Started for Glasgow Junction. Got belated in the hill woods. Inquired my way at a farmhouse and was invited to stay overnight in a rare, hearty, hospitable manner. Engaged in familiar running talk on politics, war times,

These waist-high ferns, which are silvery spleenworts (*Athyrium thelypteroides*), were growing in Daniel Boone National Forest. Muir loved ferns and was delighted to discover so many species.

and theology. The old Kentuckian seemed to take a liking to me and advised me to stay in these hills until next spring, assuring me that I would find much to interest me in and about the Great Cave; also, that he was one of the school officials and was sure that I could obtain their school for the winter term. I sincerely thanked him for his kind plans, but pursued my own.

September 7. Left the hospitable Kentuckians with their sincere good wishes and bore away southward again through the deep green woods. In noble forests all day. Saw mistletoe for the first time. . . .

September 8. . . . The scenery on approaching the Cumberland River becomes still grander. Burkesville, in beautiful location, is embosomed in a glorious array of verdant flowing hills. The Cumberland must be a happy stream. I think I could enjoy traveling with it in the midst of such beauty all my life. This evening I could find none willing to take me in, and so lay down on a hillside and fell asleep muttering praises to the happy abounding beauty of Kentucky.

September 9. Another day in the most favored province of bird and flower. Many rapid streams, flowing in beautiful flower-bordered cañons embosomed in dense woods. Am seated on a grand hill-slope that leans back against the sky like a picture. Amid the wide waves of green wood there are spots of autumnal yellow and the atmosphere, too, has the dawnings of autumn in colors and sounds. The soft light of morning falls upon ripening forests of oak and elm, walnut and hickory, and all Nature is thoughtful and calm. Kentucky is the greenest, leafiest State I have yet seen. . . .

Far the grandest of all Kentucky plants are her noble oaks. They are the master existences of her exuberant forests. Here is the Eden, the paradise of oaks. Passed the Kentucky line towards evening and obtained food and shelter from a thrifty Tennessee farmer, after he had made use of all the ordinary anti-hospitable arguments of cautious comfortable families.

Muir crossed the Cumberland River near Burkesville, but I photographed it farther east at Cumberland Falls. Man-made Lake Cumberland has destroyed a great expanse of the historic waterway, and even here at the falls, "the Niagara of the South," its water is discolored from strip-mining.

September 10. Escaped from a heap of uncordial kindness to the generous bosom of the woods. After a few miles of level ground in luxuriant tangles of brooding vines, I began the ascent of the Cumberland Mountains, the first real mountains that my foot ever touched or eyes beheld. The ascent was by a nearly regular zigzag slope, mostly covered up like a tunnel by overarching oaks. But there were a few openings where the glorious forest road of Kentucky was grandly seen, stretching over hill and valley, adjusted to every slope and curve by the hands of Nature — the most sublime and comprehensive picture that ever entered my eyes. Reached the summit in six or seven hours — a strangely long period of up-grade work to one accustomed only to the hillocky levels of Wisconsin and adjacent States.

One day I hiked to a remote part of Daniel Boone National Forest to a place called The Devil's Kitchen. There I found common wood-sorrel (*Oxalis montana*) blooming on a rock ledge by a small creek.

Muir did not pass through the Red River Gorge, in Powell County, but he did see many sandstone formations similar to those in the gorge. U.S. Forest Service ranger Don Fig and I hiked through here on a lovely autumn day. I was wearing a backpack filled with camera equipment, and at one point we climbed hand over hand up the face of a sheer rock cliff.

2
Crossing the Cumberland Mountains

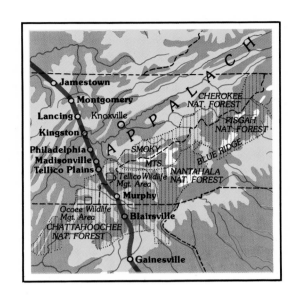

I had climbed but a short distance when I was overtaken by a young man on horseback, who soon showed that he intended to rob me if he should find the job worth while. After he had inquired where I came from, and where I was going, he offered to carry my bag. I told him that it was so light that I did not feel it at all a burden; but he insisted and coaxed until I allowed him to carry it. As soon as he had gained possession I noticed that he gradually increased his speed, evidently trying to get far enough ahead of me to examine the contents without being observed. But I was too good a walker and runner for him to get far. At a turn of the road, after trotting his horse for about half an hour, and when he thought he was out of sight, I caught him rummaging my poor bag. Finding there only a comb, brush, towel, soap, a change of underclothing, a copy of Burns's poems, Milton's Paradise Lost, and a small New Testament, he waited for me, handed back my bag, and returned down the hill, saying that he had forgotten something.

I found splendid growths of shining-leaved *Ericaceae* [heathworts] for which the Alleghany Mountains are noted. Also ferns of which *Osmunda cinnamomea* [Cinnamon Fern] is the largest and perhaps the most abundant. *Osmunda regalis* [Flowering Fern] is also common here, but not large. In Wood's and Gray's Botany *Osmunda cinnamomea* is said to be a much larger fern than *Osmunda claytoniana*. This I found to be true in Tennessee and southward, but in Indiana, part of Illinois, and Wisconsin the opposite is true. Found here the beautiful, sensitive *Schrankia*, or sensitive brier. It is a long, prickly, leguminous vine, with dense heads of small, yellow fragrant flowers.

Vines growing on roadsides receive many a tormenting blow, simply because they give evidence of feeling. Sensitive people are served in the same way. But the roadside vine soon becomes less sensitive, like people getting used to teasing — Nature, in this instance, making for the comfort of flower creatures the same benevolent arrangement as for man. Thus I found that the *Schrankia* vines growing along footpaths leading to a backwoods schoolhouse were much less sensitive than those in the adjacent unfrequented woods, having learned to pay but slight attention to the tingling strokes they get from teasing scholars.

It is startling to see the pairs of pinnate leaves rising quickly out of the grass and folding themselves close in regular succession from the root to the end of the prostrate stems, ten to twenty feet in length. How little we know as yet of the life of plants — their hopes and fears, pains and enjoyments!

The cauliflower fungus (*Sparassis spathulata*) is fairly common in Kentucky but is impressive nevertheless in form and size. Although Muir made little reference to fungi or mushrooms, which are fungi with stems, he probably passed many varieties of both.

. . . When he [a blacksmith who gave Muir shelter for the night] came in after his hard day's work and sat down to dinner, he solemnly asked a blessing on the frugal meal, consisting solely of corn bread and bacon. Then, looking across the table at me, he said, "Young man, what are you doing down here?" I replied that I was looking at plants. "Plants? What kind of plants?" I said, "Oh, all kinds; grass, weeds, flowers, trees, mosses, ferns, — almost everything that grows is interesting to me."

"Well, young man," he queried, "you mean to say that you are not employed by the Government on some private business?" "No," I said, "I am not employed by any one except just myself. I love all kinds of plants, and I came down here to these Southern States to get acquainted with as many of them as possible."

"You look like a strong-minded man," he replied, "and surely you are able to do something better than wander over the country and look at weeds and blossoms. These are hard times, and real work is required of every man that is able. Picking up blossoms doesn't seem to be a man's work at all in any kind of times."

To this I replied, "You are a believer in the Bible, are you not?" "Oh, yes." "Well, you know Solomon was a strong-minded man, and he is generally believed to have been the very wisest man the world ever saw, and yet he considered it was worth while to study plants; not only to go and pick them up as I am doing, but to study them; and you know we are told that he wrote a book about plants, not only of the great cedars of Lebanon, but of little bits of things growing in the cracks of the walls.

"Therefore, you see that Solomon differed very much more from you than from me in this matter. I'll warrant you he had many a long ramble in the mountains of Judea, and had he been a Yankee he would likely have visited every weed in the land. And again, do you not remember that Christ told his disciples to 'consider the lilies how they grow,' and compared their beauty with Solomon in all his glory? Now, whose advice am I to take, yours or Christ's? Christ says, 'Consider the lilies.' You say, 'Don't consider them. It isn't worth while for any strong-minded man.' "

Here is a whole family of little water-falls, rushing and splashing down a rock cliff. They form a part of the erosive force that was and is the major process in shaping the Smokies landscape.

This evidently satisfied him, and he acknowledged that he had never thought of blossoms in that way before. He repeated again and again that I must be a very strong-minded man, and admitted that no doubt I was fully justified in picking up blossoms. He then told me that although the war was over, walking across the Cumberland Mountains still was far from safe on account of small bands of guerrillas who were in hiding along the roads, and earnestly entreated me to turn back and not to think of walking so far as the Gulf of Mexico until the country became quiet and orderly once more.

Muir's path ran south of what is now Great Smoky Mountains National Park so he did not have the pleasure of seeing Rainbow Falls, but he surely would have approved of it. I met an eighty-one-year-old woman at Rainbow Falls who was about to make her fiftieth hike to the top of Mount Le Conte—I think Muir would have approved of her, too!

I replied that I had no fear, that I had but very little to lose, and that nobody was likely to think it worth while to rob me; that, anyhow, I always had good luck. In the morning he repeated the warning and entreated me to turn back, which never for a moment interfered with my resolution to pursue my glorious walk.

September 11. Long stretch of level sandstone plateau, lightly furrowed and dimpled with shallow groove-like valleys and hills. The trees are mostly oaks, planted wide apart like those in the Wisconsin woods. A good many pine trees here and there, forty to eighty feet high, and most of the ground is covered with showy flowers. Polygalas [milkworts], solidagoes [goldenrods], and asters were especially abundant. I came to a cool clear brook every half mile or so, the banks planted with *Osmunda regalis, Osmunda cinnamomea*, and handsome sedges. The few larger streams were fringed with laurels and azaleas. Large areas beneath the trees are

Streams of all types and sizes flow down the series of ridges and valleys that make up the state of Tennessee. The one behind this clump of trees was cold, fast, and clear. Muir must have forded many just like it.

Some lovely, serene farmland, typical of the pastoral country Muir passed through west of the Appalachians. Today agriculture is still prevalent in much of Kentucky.

47

covered with formidable green briers and brambles, armed with hooked claws, and almost impenetrable. Houses are far apart and uninhabited, orchards and fences in ruins — sad marks of war.

About noon my road became dim and at last vanished among desolate fields. Lost and hungry, I knew my direction but could not keep it on account of the briers. My path was indeed strewn with flowers, but as thorny, also, as mortal ever trod. In trying to force a way through these cat-plants one is not simply clawed and pricked through all one's clothing, but caught and held fast. The toothed arching branches come down over and above you like cruel living arms, and the more you struggle the more desperately you are entangled, and your wounds deepened and multiplied. The South has plant fly-catchers. It also has plant man-catchers. . . .

This northern maidenhair fern (*Adiantum pedatum*) seemed to be floating in mid-air above a rich base of umbrella-leaf (*Diphylleia cymosa*) when I spotted it in Great Smoky Mountains National Park.

Muir saw ferns in Tennessee as well as Kentucky. The crested fern (*Dryopteris cristata*) is a favorite of mine.

September 12. Awoke drenched with mountain mist, which made a grand show, as it moved away before the hot sun. Passed Montgomery, a shabby village at the head of the east slope of the Cumberland Mountains. Obtained breakfast in a clean house and began the descent of the mountains. Obtained fine views of a wide, open country, and distant flanking ridges and spurs. Crossed a wide cool stream [Emory River], a branch of the Clinch River. There is nothing more eloquent in Nature than a mountain stream, and this is the first I ever saw. Its banks are luxuriantly peopled with rare and lovely flowers and overarching trees, making one of Nature's coolest and most hospitable places. Every tree, every flower, every ripple and eddy of this lovely stream seemed solemnly to feel the presence of the great Creator. Lingered in this sanctuary a long time thanking the Lord with all my heart for his goodness in allowing me to enter and enjoy it.

Discovered two ferns, *Dicksonia* and a small matted polypod on trees, common farther South. Also a species of magnolia with very large leaves and scarlet conical fruit. Near this stream I spent some joyous time in a grand rock-dwelling full of mosses, birds, and flowers. Most heavenly place I ever entered. The long narrow valleys of the mountainside, all well watered and nobly adorned with oaks, magnolias, laurels, azaleas, asters, ferns, Hypnum mosses, Madotheca [Scale-mosses], etc. Also towering clumps of beautiful hemlocks. The hemlock, judging from the common species of Canada, I regarded as the least noble of the conifers. But those of the eastern valleys of the Cumberland Mountains are as perfect in form and regal in port as the pines themselves. The latter abundant. Obtained fine glimpses from open places as I descended to the great valley between these mountains and the Unaka Mountains on the state line. Forded the Clinch, a beautiful clear stream, that knows many of the dearest mountain retreats

that ever heard the music of running water. Reached Kingston before dark. Sent back my plant collections by express to my brother in Wisconsin. . . .

September 15. Most glorious billowy mountain scenery. Made many a halt at open places to take breath and to admire. The road, in many places cut into the rock, goes winding about among the knobs and gorges. Dense growth of asters, liatris, and grapevines.

Reached a house before night, and asked leave to stop. "Well, you're welcome to stop," said the mountaineer, "if you think you can live till morning on what I have to live on all the time." Found the old gentleman very communicative. Was favored with long "bar" stories, deer hunts, etc., and in the morning was pressed to stay a day or two.

Seeing a waterfall is as relaxing for me as wading barefoot in ocean surf. Abrams Falls in Great Smoky Mountains National Park must have the same effect on other people also because every year thousands of tourists leave their cars behind and walk the two and a half miles to visit it.

September 16. "I will take you," said he, "to the highest ridge in the country, where you can see both ways. You will have a view of all the world on one side of the mountains and all creation on the other. Besides, you, who are traveling for curiosity and wonder, ought to see our gold mines." I agreed to stay and went to the mines. Gold is found in small quantities throughout the Alleghanies, and many farmers work at mining a few weeks or months every year when their time is not more valuable for other pursuits. In this neighborhood miners are earning from half a dollar to two dollars a day. There are several large quartz mills not far from here. Common labor is worth ten dollars a month.

September 17. Spent the day in botanizing, blacksmithing, and examining a grist mill. Grist mills, in the less settled parts of Tennessee and North Carolina, are remarkably simple affairs. A small stone, that a man might carry under his arm, is fastened to the vertical shaft of a little home-made, boyish-looking, back-action water-wheel, which, with a hopper and a box

In Great Smoky Mountains National Park by the West Prong of the Little Pigeon River, early one morning — I stubbed my toe badly. But I'm inclined to think this back-lit photograph featuring the morning haze was worth it.

Mingus Mill was built in the Oconaluftee River valley (now in the Smokies park) by an immigrant from Germany in the late 1790s. It is probably similar to the gristmills that Muir wrote about. The National Park Service has restored the mill, and it is open to the public.

To me there is nothing as enchanting as traveling on a trail by a river through rain or dense fog. I found this magical scene in Shining Rock Wilderness in Pisgah National Forest. Shining Rock was one of the two earliest Forest Service Wilderness areas established in the Southeast.

to receive the meal, is the whole affair. The walls of the mill are of undressed poles cut from seedling trees and there is no floor, as lumber is dear. No dam is built. The water is conveyed along some hillside until sufficient fall is obtained, a thing easily done in the mountains.

On Sundays you may see wild, unshorn, uncombed men coming out of the woods, each with a bag of corn on his back. From a peck to a bushel is a common grist. They go to the mill along verdant footpaths, winding up and down over hill and valley, and crossing many a rhododendron glen. The flowers and shining leaves brush against their shoulders and knees, occasionally knocking off their coon-skin caps. The first arrived throws his corn into the hopper, turns on the water, and goes to the house. After chatting and smoking he returns to see if his grist is done. Should the stones run empty for an hour or two, it does no harm. . . .

When every second counts, it seems, I'm frequently out of film. Running for more so I could capture this Smokies sunset, I fell and fractured my skull. The accident was a result of my usual photographic method: not planning pictures ahead of time but relying instead on instinct and discovery.

September 18. Up the mountain on the state line. The scenery is far grander than any I ever before beheld. The view extends from the Cumberland Mountains on the north far into Georgia and North Carolina to the south, an area of about five thousand square miles. Such an ocean of wooded, waving, swelling mountain beauty and grandeur is not to be

described. Countless forest-clad hills, side by side in rows and groups, seemed to be enjoying the rich sunshine and remaining motionless only because they were so eagerly absorbing it. All were united by curves and slopes of inimitable softness and beauty. Oh, these forest gardens of our Father! What perfection, what divinity, in their architecture! What simplicity and mysterious complexity of detail! Who shall read the teaching of these sylvan pages, the glad brotherhood of rills that sing in the valleys, and all the happy creatures that dwell in them under the tender keeping of a Father's care?

"Track Gap," the wonderful pass that a mountaineer described to Muir, was most certainly Track Rock Gap near Blairsville, Georgia. I photographed this Indian carving of a turkey's foot on one of its boulders.

September 19. Received another solemn warning of dangers on my way through the mountains. Was told by my worthy entertainer of a wondrous gap in the mountains which he advised me to see. "It is called Track Gap," said he, "from the great number of tracks in the rocks — bird tracks, bar tracks, hoss tracks, men tracks, all in the solid rock as if it had been mud." Bidding farewell to my worthy mountaineer and all his comfortable wonders, I pursued my way to the South.

As I was leaving, he repeated the warnings of danger ahead, saying that there were a good many people living like wild beasts on whatever they could steal, and that murders were sometimes committed for four or five dollars, and even less. While stopping with him I noticed that a man came regularly after dark to the house for his supper. He was armed with a gun, a pistol, and a long knife. My host told me that this man was at feud with one of his neighbors, and that they were prepared to shoot one another at sight. That neither of them could do any regular work or sleep in the same place two nights in succession. That they visited houses only for food, and as soon as the one that I saw had got his supper he went out and slept in the woods, without of course making a fire. His enemy did the same.

My entertainer told me that he was trying to make peace between these two men, because they both were good men, and if they would agree to stop their quarrel, they could then both go to work. Most of the food in this house was coffee without sugar, corn bread, and sometimes bacon. But the coffee was the greatest luxury which these people knew. The only way of obtaining it was by selling skins, or, in particular, "sang," that is ginseng, which found a market in far-off China.

My path all to-day led me along the leafy banks of the Hiwassee, a most impressive mountain river. Its channel is very rough, as it crosses the edges of upturned rock strata, some of them standing at right angles,

The section of the Hiwassee River that Muir followed is now drowned under a man-made impoundment, Lake Hiwassee. So I have illustrated that river instead with a photograph of the Tellico River. It has the same unusual strata of upturned rock that used to be visible along the Hiwassee.

Waucheesi Mountain in Monroe County, Tennessee, may have been the exact spot where Muir crossed into North Carolina, but I was never able to photograph the view from there because the air was always full of haze and pollution from Tennessee factories. Perhaps, in Muir's day, the view looked something like this present-day view of Linville River Gorge in Pisgah National Forest. I cracked a rib as I leaned out from the rock cliffs to take this picture.

or glancing off obliquely to right and left. Thus a multitude of short, resounding cataracts are produced, and the river is restrained from the headlong speed due to its volume and the inclination of its bed.

All the larger streams of uncultivated countries are mysteriously charming and beautiful, whether flowing in mountains or through swamps and plains. Their channels are interestingly sculptured, far more so than the grandest architectural works of man. The finest of the forests are usually found along their banks, and in the multitude of falls and rapids the wilderness finds a voice. Such a river is the Hiwassee, with its surface broken to a thousand sparkling gems, and its forest walls vine-draped and flowery as Eden. And how fine the songs it sings! . . .

September 20. All day among the groves and gorges of Murphy with Mr. Beale. Was shown the site of Camp Butler where General Scott had his headquarters when he removed the Cherokee Indians to a new home in the West. Found a number of rare and strange plants on the rocky banks of the river Hiwassee. In the afternoon, from the summit of a commanding

Pale-spike lobelia (*Lobelia spicata*) in Linville Gorge. Lobelias are in the bluebell family.

It is too bad that Muir did not *see* mountain laurel (*Kalmia latifolia*) in bloom, but perhaps he pressed some of its leaves in the plant press he was carrying.

56

ridge, I obtained a magnificent view of blue, softly curved mountain scenery. Among the trees I saw *Ilex* [Holly] for the first time. . . .

September 22. Hills becoming small, sparsely covered with soil. They are called "knob land" and are cultivated, or scratched, with a kind of one-tooth cultivator. Every rain robs them of their fertility, while the bottoms are of course correspondingly enriched. About noon I reached the last mountain summit on my way to the sea. It is called the Blue Ridge and before it lies a prospect very different from any I had passed, namely, a vast uniform expanse of dark pine woods, extending to the sea; an impressive view at any time and under any circumstances, but particularly so to one emerging from the mountains.

        Traveled in the wake of three poor but merry mountaineers — an old woman, a young woman, and a young man — who sat, leaned, and lay in the box of a shackly wagon that seemed to be held together by spiritualism, and was kept in agitation by a very large and a very small mule. In going down hill the looseness of the harness and the joints of the wagon allowed

On Muir's final evening in Tennessee, he may have seen such a mountain sunset. U.S. Forest Service ranger Harvey Price, a staunch Muir disciple, believes that the following day the naturalist crossed the Unicoi Mountains into North Carolina via the Warriors' Trail, which was made by the English in 1760 to send supplies from Charleston, South Carolina, to a fort in the Tellico valley. The path has been grown over for years, but a portion was cleared recently by boy scouts.

Muir was impressed by the towering hemlocks in the southern Appalachians. Here, in Joyce Kilmer Memorial Forest, they still grow in great numbers along with a stand of virgin tulip-poplars, perhaps the last stand of its size in the country.

the mules to back nearly out of sight beneath the box, and the three who occupied it were slid against the front boards in a heap over the mules' ears. Before they could unravel their limbs from this unmannerly and impolite disorder, a new ridge in the road frequently tilted them with a swish and a bump against the back boards in a mixing that was still more grotesque.

I expected to see man, women, and mules mingled in piebald ruin at the bottom of some rocky hollow, but they seemed to have full confidence in the back board and front board of the wagon-box. So they continued to slide comfortably up and down, from end to end, in slippery obedience to the law of gravitation, as the grades demanded. Where the jolting was moderate, they engaged in conversation on love, marriage, and camp-meeting, according to the custom of the country. The old lady, through all the vicissitudes of the transportation, held a bouquet of French marigolds.

The hillsides hereabouts were bearing a fine harvest of asters. Reached Mount Yonah in the evening. Had a long conversation with an old Methodist slaveholder and mine owner. Was hospitably refreshed with a drink of fine cider.

3
Through the
River Country of Georgia

September 23. Am now fairly out of the mountains. Thus far the climate has not changed in any marked degree, the decrease in latitude being balanced by the increase in altitude. These mountains are highways on which northern plants may extend their colonies southward. The plants of the North and of the South have many minor places of meeting along the way I have traveled; but it is here on the southern slope of the Alleghanies that the greatest number of hardy, enterprising representatives of the two climates are assembled.

Passed the comfortable, finely shaded little town of Gainesville. The Chattahoochee River is richly embanked with massive, bossy, dark green water oaks, and wreathed with a dense growth of muscadine grapevines, whose ornate foliage, so well adapted to bank embroidery, was enriched with other interweaving species of vines and brightly colored flowers. This is the first truly southern stream I have met.

At night I reached the home of a young man with whom I had worked in Indiana, Mr. Prater. He was down here on a visit to his father and mother. This was a plain backwoods family, living out of sight among knobby timbered hillocks not far from the river. The evening was passed in mixed conversation on southern and northern generalities.

September 24. Spent this day with Mr. Prater sailing on the Chattahoochee, feasting on grapes that had dropped from the overhanging

vines. This remarkable species of wild grape has a stout stem, sometimes five or six inches in diameter, smooth bark and hard wood, quite unlike any other wild or cultivated grapevine that I have seen. The grapes are very large, some of them nearly an inch in diameter, globular and fine flavored. Usually there are but three or four berries in a cluster, and when mature they drop off instead of decaying on the vine. Those which fall into the river are often found in large quantities in the eddies along the bank, where they are collected by men in boats and sometimes made into wine. I think another name for this grape is the Scuppernong [the southern species of foxgrape, *Vitis rotundifolia*], though called ''muscadine'' here.

I feel closer to John Muir in Linville Gorge than anywhere else because soon after taking this photograph of gathering storm clouds I was caught in the magnificent storm I described in the preface of this book.

Besides sailing on the river, we had a long walk among the plant bowers and tangles of the Chattahoochee bottom lands.

September 25. Bade good-bye to this friendly family. Mr. Prater accompanied me a short distance from the house and warned me over and over again to be on the outlook for rattlesnakes. They are now leaving the damp lowlands, he told me, so that the danger is much greater because they are on their travels. Thus warned, I set out for Savannah, but got lost in the vine-fenced hills and hollows of the river bottom. Was unable to find the ford to which I had been directed by Mr. Prater.

I then determined to push on southward regardless of roads and fords. After repeated failures I succeeded in finding a place on the river bank where I could force my way into the stream through the vine-tangles. I succeeded in crossing the river by wading and swimming, careless of wetting, knowing that I would soon dry in the hot sunshine.

Out near the middle of the river I found great difficulty in resisting the rapid current. Though I braced myself with a stout stick, I was at length carried away in spite of all my efforts. But I succeeded in swimming to the shallows on the farther side, luckily caught hold of a rock, and after a rest swam and waded ashore. Dragging myself up the steep bank by the overhanging vines, I spread out myself, my paper money, and my plants to dry.

Debated with myself whether to proceed down the river valley until I could buy a boat, or lumber to make one, for a sail instead of a march through Georgia. I was intoxicated with the beauty of these glorious river banks, which I fancied might increase in grandeur as I approached the sea. But I finally concluded that such a pleasure sail would be less profitable than a walk, and so sauntered on southward as soon as I was dry. Rattlesnakes abundant. Lodged at a farmhouse. Found a few tropical plants in the garden.

Cotton is the principal crop hereabouts, and picking is now going on merrily. Only the lower bolls are now ripe. Those higher on the plants are green and unopened. Higher still, there are buds and flowers, some of which, if the plants be thrifty and the season favorable, will continue to produce ripe bolls until January. . . .

September 26. Reached Athens in the afternoon, a remarkably beautiful and aristocratic town, containing many classic and magnificent mansions of wealthy planters, who formerly owned large negro-stocked plantations in the best cotton and sugar regions farther south. Unmistakable marks of culture and refinement, as well as wealth, were everywhere apparent. This is the most beautiful town I have seen on the journey, so far, and the only one in the South that I would like to revisit. . . .

I have lived most of my life in central Georgia, but until I traced John Muir's walk, I had never stopped to consider the butterfly pea (*Clitoria mariana*). This beautifully shaped flower measured two inches across.

Fall leaves floating on the Linville River.

Muir would have observed a large variety of wildflowers in bloom, such as this spotted touch-me-not (*Impatiens capensis*), had he made his famous walk in April and May. Flower lovers return to Great Smoky Mountains National Park year after year during wildflower season.

I made this photograph so quickly, in order to catch the back-lighting of the turk's-cap lily (*Lilium superbum*), that I failed to notice the ginseng (*Panax quinquefolius*) in the lower left. Later, while viewing the slide, I realized I had photographed two unusual plants instead of one.

Wild potato-vine or man-of-the-earth (*Ipomoea pandurata*), photographed here in central Georgia, belongs to the morning-glory family. Its bell-like flowers are large—two or three inches wide.

The catawba rhododendrons (*Rhododendron catawbiense*) in Linville Gorge are the most magnificent I've ever seen. There are several species growing here that are native only to the gorge.

September 27. When very thirsty I discovered a beautiful spring in a sandstone basin overhung with shady bushes and vines, where I enjoyed to the utmost the blessing of pure cold water. Discovered here a fine southern fern, some new grasses, etc. Fancied that I might have been directed here by Providence, while fainting with thirst. It is not often hereabouts that the joys of cool water, cool shade, and rare plants are so delightfully combined.

Witnessed the most gorgeous sunset I ever enjoyed in this bright world of light. The sunny South is indeed sunny. . . .

September 28. The water oak is abundant on stream banks and in damp hollows. Grasses are becoming tall and cane-like and do not cover the ground with their leaves as at the North. Strange plants are crowding about me now. Scarce a familiar face appears among all the flowers of the day's walk.

September 29. To-day I met a magnificent grass, ten or twelve feet in stature, with a superb panicle of glossy purple flowers. Its leaves, too, are of

princely mould and dimensions. Its home is in sunny meadows and along the wet borders of slow streams and swamps. It seems to be fully aware of its high rank, and waves with the grace and solemn majesty of a mountain pine. I wish I could place one of these regal plants among the grass settlements of our Western prairies. Surely every panicle would wave and bow in joyous allegiance and acknowledge their king.

September 30. Between Thomson and Augusta I found many new and beautiful grasses, tall gerardias, liatris, club mosses, etc. Here, too, is the northern limit of the remarkable long-leafed pine, a tree from sixty to seventy feet in height, from twenty to thirty inches in diameter, with leaves ten to fifteen inches long, in dense radiant masses at the ends of the naked branches. The wood is strong, hard, and very resinous. It makes excellent ship spars, bridge timbers, and flooring. Much of it is shipped to the West India Islands, New York, and Galveston.

The seedlings, five or six years old, are very striking objects to one from the North, consisting, as they do, of the straight, leafless stem, surmounted by a crown of deep green leaves, arching and spreading like a palm. Children fancy that they resemble brooms, and use them as such in their picnic play-houses. *Pinus palustris* is most abundant in Georgia and Florida.

The sandy soil here is sparingly seamed with rolled quartz pebbles and clay. Denudation, going on slowly, allows the thorough removal of these clay seams, leaving only the sand. Notwithstanding the sandiness of the soil, much of the surface of the country is covered with standing water, which is easily accounted for by the presence of the above-mentioned impermeable seams.

Traveled to-day more than forty miles without dinner or supper. No family would receive me, so I had to push on to Augusta. Went hungry to bed and awoke with a sore stomach — sore, I suppose, from its walls rubbing on each other without anything to grind. A negro kindly directed me to the best hotel, called, I think, the Planter's. Got a good bed for a dollar.

October 1. Found a cheap breakfast in a market-place; then set off along the Savannah River to Savannah. Splendid grasses and rich, dense, vine-clad forests. Muscadine grapes in cart-loads. Asters and solidagoes becom-

Most of the impenetrable cypress swamps that Muir saw along the Savannah River no longer exist, so I was grateful to be able to photograph the pond cypresses (*Taxodium ascendens*) in Okefenokee National Wildlife Refuge. Only a few virgin cypress trees still grow in Georgia and South Carolina.

ing scarce. Carices [sedges] quite rare. Leguminous plants abundant. A species of passion flower is common, reaching back into Tennessee. It is here called "apricot vine," has a superb flower, and the most delicious fruit I have ever eaten.

The pomegranate is cultivated here. The fruit is about the size of an orange, has a thick, tough skin, and when opened resembles a many-chambered box full of translucent purple candies.

Toward evening I came to the country of one of the most striking of southern plants, the so-called "Long Moss" or Spanish Moss [Tillandsia], though it is a flowering plant and belongs to the same family as the pineapple [Bromelworts]. The trees hereabouts have all their branches draped with it, producing a remarkable effect.

Here, too, I found an impenetrable cypress swamp. This remarkable tree, called cypress, is a taxodium, grows large and high, and is remarkable for its flat crown. The whole forest seems almost level on the top, as if each tree had grown up against a ceiling, or had been rolled while growing. This taxodium is the only level-topped tree that I have seen. The branches, though spreading, are careful not to pass each other, and stop suddenly on reaching the general level, as if they had grown up against a ceiling.

The groves and thickets of smaller trees are full of blooming evergreen vines. These vines are not arranged in separate groups, or in delicate wreaths, but in bossy walls and heavy, mound-like heaps and banks. Am made to feel that I am now in a strange land. I know hardly any of the plants, but few of the birds, and I am unable to see the country for the solemn, dark, mysterious cypress woods which cover everything.

The winds are full of strange sounds, making one feel far from the people and plants and fruitful fields of home. Night is coming on and I am filled with indescribable loneliness. Felt feverish; bathed in a black, silent stream; nervously watchful for alligators. Obtained lodging in a planter's house among cotton fields. Although the family seemed to be pretty well-off, the only light in the house was bits of pitch-pine wood burned in the fireplace.

October 2. In the low bottom forest of the Savannah River. Very busy with new specimens. Most exquisitely planned wrecks of *Agrostis scabra* [Rough Hair Grass]. Pines in glorious array with open, welcoming, approachable plants. . . .

The wood ibis (*Mycteria americana*) and numerous other species of waterbirds can be observed in Okefenokee Swamp. Fortunately, the refuge has recently been designated a part of the National Wilderness Preservation System.

October 3. In "pine barrens" most of the day. Low, level, sandy tracts; the pines wide apart; the sunny spaces between full of beautiful abounding grasses, liatris, long, wand-like solidago, saw palmettos, etc., covering the ground in garden style. Here I sauntered in delightful freedom, meeting none of the cat-clawed vines, or shrubs, of the alluvial bottoms. Dwarf live-oaks common. . . .

October 5. Saw the stately banana for the first time, growing luxuriantly in the wayside gardens. At night with a very pleasant, intelligent Savannah family, but as usual was admitted only after I had undergone a severe course of questioning.

October 6. Immense swamps, still more completely fenced and darkened, that are never ruffled with winds or scorched with drought. Many of them seem to be thoroughly aquatic.

October 7. Impenetrable taxodium swamp, seemingly boundless. The silvery skeins of tillandsia becoming longer and more abundant. Passed the night with a very pleasant family of Georgians, after the usual questions and cross questions.

October 8. Found the first woody *compositae*, a most notable discovery. Took them to be such at a considerable distance. Almost all trees and shrubs are evergreens here with thick polished leaves. *Magnolia grandiflora* becoming common. A magnificent tree in fruit and foliage as well as in flower. Near Savannah I found waste places covered with a dense growth of woody leguminous plants, eight or ten feet high, with pinnate leaves and suspended rattling pods.

Reached Savannah, but find no word from home, and the money that I had ordered to be sent by express from Portage [Wisconsin] by my brother had not yet arrived. Feel dreadfully lonesome and poor. Went to the meanest looking lodging-house that I could find, on account of its cheapness.

Muir wrote in his journal, "Of the people of the states that I have now passed, I best like the Georgians." He must have also liked Georgia's soft little waterfalls, such as Henry's Rainbow Falls in Chattahoochee National Forest.

4
Camping Among the Tombs

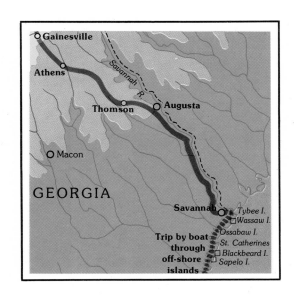

October 9. After going again to the express office and post office, and wandering about the streets, I found a road which led me to the Bonaventure graveyard. If that burying-ground across the Sea of Galilee, mentioned in Scripture, was half as beautiful as Bonaventure, I do not wonder that a man should dwell among the tombs. It is only three or four miles from Savannah, and is reached by a smooth white shell road.

There is but little to be seen on the way in land, water, or sky, that would lead one to hope for the glories of Bonaventure. The ragged desolate fields, on both sides of the road, are overrun with coarse rank weeds, and show scarce a trace of cultivation. But soon all is changed. Rickety log huts, broken fences, and that last patch of weedy rice-stubble are left behind. You come to beds of purple liatris and living wild-wood trees. You hear the song of birds, cross a small stream, and are with Nature in the grand old forest graveyard, so beautiful that almost any sensible person would choose to dwell here with the dead rather than with the lazy, disorderly living.

Part of the grounds was cultivated and planted with live-oak, about a hundred years ago, by a wealthy gentleman who had his country residence here. But much the greater part is undisturbed. Even those spots which are disordered by art, Nature is ever at work to reclaim, and to make them look as if the foot of man had never known them. Only a small plot of ground is occupied with graves and the old mansion is in ruins.

77

I visited Savannah's Bonaventure
Cemetery several times and walked
from tomb to tomb, trying to determine the location of Muir's "nest" in
a sparkleberry thicket. Many of the
magnificent live-oaks that impressed
him are still growing here, but the old
mansion is gone.

The most conspicuous glory of Bonaventure is its noble avenue of
live-oaks. They are the most magnificent planted trees I have ever seen,
about fifty feet high and perhaps three or four feet in diameter, with broad
spreading leafy heads. The main branches reach out horizontally until they
come together over the driveway, embowering it throughout its entire
length, while each branch is adorned like a garden with ferns, flowers,
grasses, and dwarf palmettos.

But of all the plants of these curious tree-gardens the most striking and
characteristic is the so-called Long Moss (*Tillandsia usneoides*). It drapes all
the branches from top to bottom, hanging in long silvery-gray skeins,
reaching a length of not less than eight or ten feet, and when slowly waving
in the wind they produce a solemn funereal effect singularly impressive.

There are also thousands of smaller trees and clustered bushes, covered almost from sight in the glorious brightness of their own light. The
place is half surrounded by the salt marshes and islands of the river, their
reeds and sedges making a delightful fringe. Many bald eagles roost among

the trees along the side of the marsh. Their screams are heard every morning, joined with the noise of crows and the songs of countless warblers, hidden deep in their dwellings of leafy bowers. Large flocks of butterflies, all kinds of happy insects, seem to be in a perfect fever of joy and sportive gladness. The whole place seems like a center of life. The dead do not reign there alone.

Bonaventure to me is one of the most impressive assemblages of animal and plant creatures I ever met. I was fresh from the Western prairies, the garden-like openings of Wisconsin, the beech and maple and oak woods of Indiana and Kentucky, the dark mysterious Savannah cypress forests; but never since I was allowed to walk the woods have I found so impressive a company of trees as the tillandsia-draped oaks of Bonaventure.

I gazed awe-stricken as one new-arrived from another world. Bonaventure is called a graveyard, a town of the dead, but the few graves are powerless in such a depth of life. The rippling of living waters, the song

Today Muir would not recognize the view of the salt marsh he saw from Bonaventure Cemetery because a bridge and causeway cut the marsh in two. But this marsh on nearby Ossabaw Island probably looks much like his pre-bridge graveyard view.

79

Caesar's amanita (*Amanita caesarea*) tastes as good as it looks. But because it's an amanita and other poisonous amanitas resemble it, positive identification is a must before eating. (Central Georgia)

The fly amanita (*Amanita muscaria*) is a most attractive mushroom and can appear in any of three or four different colors. It is poisonous but not deadly. (Central Georgia)

The death cap mushroom (*Amanita phalloides*) is deadly poisonous. It is one of three amanitas said to account for 90 percent of all deaths from mushroom poisoning. (Central Georgia)

of birds, the joyous confidence of flowers, the calm, undisturbable grandeur of the oaks, mark this place of graves as one of the Lord's most favored abodes of life and light. . . .

Most of the few graves of Bonaventure are planted with flowers. There is generally a magnolia at the head, near the strictly erect marble, a rose-bush or two at the foot, and some violets and showy exotics along the sides or on the tops. All is enclosed by a black iron railing, composed of rigid bars that might have been spears or bludgeons from a battlefield in Pandemonium.

It is interesting to observe how assiduously Nature seeks to remedy these labored art blunders. She corrodes the iron and marble, and gradually levels the hill which is always heaped up, as if a sufficiently heavy quantity of clods could not be laid on the dead. Arching grasses come one by one; seeds come flying on downy wings, silent as fate, to give life's dearest beauty for the ashes of art; and strong evergreen arms laden with ferns and tillandsia drapery are spread over all — Life at work everywhere, obliterating all memory of the confusion of man. . . .

A spider bit me on the nose as I lay on the ground photographing this purple russula (*Russula purpurina*) in Chattahoochee
National Forest. But I was the intruder.

The panther amanita or killer cat mushroom (*Amanita
pantherina*), as its name implies, is poisonous. This specimen
was growing in Chattahoochee National Forest.

The money package that I was expecting did not arrive until the
following week. After stopping the first night at the cheap, disreputable-
looking hotel, I had only about a dollar and a half left in my purse, and so
was compelled to camp out to make it last in buying only bread. . . .

By this time it was near sunset, and I hastened across the common to
the road and set off for Bonaventure, delighted with my choice, and almost
glad to find that necessity had furnished me with so good an excuse for
doing what I knew my mother would censure; for she made me promise I
would not lie out of doors if I could possibly avoid it. The sun was set ere I
was past the negroes' huts and rice fields, and I arrived near the graves in
the silent hour of the gloaming.

I was very thirsty after walking so long in the muggy heat, a distance of
three or four miles from the city, to get to this graveyard. A dull, sluggish,
coffee-colored stream flows under the road just outside the graveyard
garden park, from which I managed to get a drink after breaking a way
down to the water through a dense fringe of bushes, daring the snakes and
alligators in the dark. Thus refreshed I entered the weird and beautiful
abode of the dead.

All the avenue where I walked was in shadow, but an exposed tombstone frequently shone out in startling whiteness on either hand, and thickets of sparkleberry bushes gleamed like heaps of crystals. Not a breath of air moved the gray moss, and the great black arms of the trees met overhead and covered the avenue. But the canopy was fissured by many a netted seam and leafy-edged opening, through which the moonlight sifted in auroral rays, broidering the blackness in silvery light. Though tired, I sauntered a while enchanted, then lay down under one of the great oaks. I found a little mound that served for a pillow, placed my plant press and bag beside me and rested fairly well, though somewhat disturbed by large prickly-footed beetles creeping across my hands and face, and by a lot of hungry stinging mosquitoes.

When I awoke, the sun was up and Nature was rejoicing. Some birds had discovered me as an intruder, and were making a great ado in interesting language and gestures. I heard the screaming of the bald eagles, and of some strange waders in the rushes. I heard the hum of Savannah with the long jarring hallos of negroes far away. On rising I found that my head had been resting on a grave, and though my sleep had not been quite so sound as that of the person below, I arose refreshed, and looking about me, the morning sunbeams pouring through the oaks and gardens dripping with dew, the beauty displayed was so glorious and exhilarating that hunger and care seemed only a dream.

Eating a breakfast cracker or two and watching for a few hours the beautiful light, birds, squirrels, and insects, I returned to Savannah, to find that my money package had not yet arrived. I then decided to go early to the graveyard and make a nest with a roof to keep off the dew, as there was no way of finding out how long I might have to stay. I chose a hidden spot in a dense thicket of sparkleberry bushes, near the right bank of the Savannah River, where the bald eagles and a multitude of singing birds roosted. It was so well hidden that I had to carefully fix its compass bearing in my mind from a mark I made on the side of the main avenue, that I might be able to find it at bedtime.

I used four of the bushes as corner posts for my little hut, which was about four or five feet long by about three or four in width, tied little branches across from forks in the bushes to support a roof of rushes, and spread a thick mattress of Long Moss over the floor for a bed. My whole establishment was on so small a scale that I could have taken up, not only my bed, but my whole house, and walked. There I lay that night, eating a few crackers.

Next day I returned to the town and was disappointed as usual in obtaining money. So after spending the day looking at the plants in the gardens of the fine residences and town squares, I returned to my graveyard home. That I might not be observed and suspected of hiding, as if I had committed a crime, I always went home after dark, and one night, as I lay down in my moss nest, I felt some cold-blooded creature in it; whether a snake or simply a frog or toad I do not know, but instinctively, instead of drawing back my hand, I grasped the poor creature and threw it over the tops of the bushes. That was the only significant disturbance or fright that I got.

In the morning everything seemed divine. Only squirrels, sunbeams, and birds came about me. I was awakened every morning by these little singers after they discovered my nest. Instead of serenely singing their morning songs they at first came within two or three feet of the hut, and, looking in at me through the leaves, chattered and scolded in half-angry, half-wondering tones. The crowd constantly increased, attracted by the disturbance. Thus I began to get acquainted with my bird neighbors in this blessed wilderness, and after they learned that I meant them no ill they scolded less and sang more.

After five days of this graveyard life I saw that even with living on three or four cents a day my last twenty-five cents would soon be spent, and after trying again and again unsuccessfully to find some employment began to think that I must strike farther out into the country, but still within reach of town, until I came to some grain or rice field that had not yet been harvested, trusting that I could live indefinitely on toasted or raw corn, or rice.

By this time I was becoming faint, and in making the journey to the town was alarmed to find myself growing staggery and giddy. The ground ahead seemed to be rising up in front of me, and the little streams in the ditches on the sides of the road seemed to be flowing up hill. Then I realized that I was becoming dangerously hungry and became more than ever anxious to receive that money package.

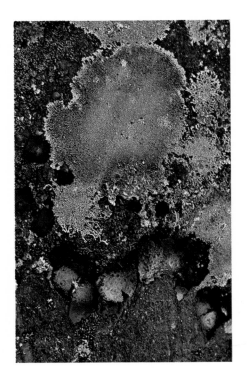

Too many of us completely miss the joy of discovering magnificent forms and colors of small plants. So it is with lichens, which are part fungus and part alga. I had trouble identifying this rich blue growth at the bottom of Linville Gorge. Lichenology needs more lichenologists.

To my delight this fifth or sixth morning, when I inquired if the money package had come, the clerk replied that it had, but that he could not deliver it without my being identified. I said, "Well, here! read my brother's letter," handing it to him. "It states the amount in the package, where it came from, the day it was put into the office at Portage City, and I should think that would be enough." He said, "No, that is not enough. How do I know that this letter is yours? You may have stolen it. How do I know that you are John Muir?"

I said, "Well, don't you see that this letter indicates that I am a botanist? For in it my brother says, 'I hope you are having a good time and finding many new plants.' Now, you say that I might have stolen this letter from John Muir, and in that way have become aware of there being a money package to arrive from Portage for him. But the letter proves that John Muir must be a botanist, and though, as you say, his letter might have been stolen, it would hardly be likely that the robber would be able to steal John Muir's knowledge of botany. Now I suppose, of course, that you have been to school and know something of botany. Examine me and see if I know anything about it."

At this he laughed good-naturedly, evidently feeling the force of my argument, and, perhaps, pitying me on account of looking pale and hungry, he turned and rapped at the door of a private office — probably the Manager's — called him out and said, "Mr. So and So, here is a man who has inquired every day for the last week or so for a money package from Portage, Wisconsin. He is a stranger in the city with no one to identify him. He states correctly the amount and the name of the sender. He has shown me a letter which indicates that Mr. Muir is a botanist, and that although a traveling companion may have stolen Mr. Muir's letter, he could not have stolen his botany, and requests us to examine him."

The head official smiled, took a good stare into my face, waved his hand, and said, "Let him have it." Gladly I pocketed my money, and had not gone along the street more than a few rods before I met a very large negro woman with a tray of gingerbread, in which I immediately invested some of my new wealth, and walked rejoicingly, munching along the street, making no attempt to conceal the pleasure I had in eating. Then, still hunting for more food, I found a sort of eating-place in a market and had a large regular meal on top of the gingerbread! Thus my "marching through Georgia" terminated handsomely in a jubilee of bread.

The edge of a marsh on Ossabaw Island. Strangely, in 1867 most of the Georgia islands didn't look very wild: they were suffering from extensive farming and timbering. It has taken them a century to lose the scars of pre-Civil War cotton plantations, some of which were worked by thousands of slaves.

Many pieces of dead cedar, twisted like taffy, lie scattered on Stone Mountain, which is now a state park. Muir missed this natural wonder by only a few miles. Had he known it was there, he would surely have detoured to climb it.

When I took this photograph of a dead tree in a tidal pond, I imagined that Muir might have seen the same beach from shipboard—if only briefly.

5
Through Florida Swamps and Forests

... Since the commencement of my floral pilgrimage I have seen much that is not only new, but altogether unallied, unacquainted with the plants of my former life. I have seen magnolias, tupelo, live-oak, Kentucky oak, tillandsia, long-leafed pine, palmetto, schrankia, and whole forests of strange trees and vine-tied thickets of blooming shrubs; whole meadowfuls of magnificent bamboo and lakefuls of lilies, all new to me; yet I still press eagerly on to Florida as the special home of the tropical plants I am looking for, and I feel sure I shall not be disappointed.

The same day on which the money arrived I took passage on the steamship Sylvan Shore for Fernandina, Florida. The daylight part of this sail along the coast of Florida was full of novelty, and by association awakened memories of my Scottish days at Dunbar on the Firth of Forth.... Altogether my half-day and night on board the steamer were pleasant, and carried me past a very sickly, entangled, overflowed, and unwalkable piece of forest.

It is pretty well known that a short geological time ago the ocean covered the sandy level margin, extending from the foot of the Alleghanies to the present coast-line, and in receding left many basins for lakes and swamps. The land is still encroaching on the sea, and it does so not evenly, in a regular line, but in fringing lagoons and inlets and dotlike coral islands.

Although Muir wrote that the daylight part of his sail after leaving Savannah was along the coast of Florida, he was almost certainly mistaken. The *Sylvan Shore* would have cruised along the coast of Georgia and passed all the Georgia sea islands, including Cumberland, where I captured this sunset.

It is on the coast strip of isles and peninsulas that sea-island cotton is grown. Some of these small islands are afloat, anchored only by the roots of mangroves and rushes. For a few hours our steamer sailed in the open sea, exposed to its waves, but most of the time she threaded her way among the lagoons, the home of alligators and countless ducks and waders.

October 15. To-day, at last, I reached Florida, the so-called "Land of Flowers," that I had so long waited for, wondering if after all my longings and prayers would be in vain, and I should die without a glimpse of the flowery Canaan. But here it is, at the distance of a few yards! — a flat, watery, reedy coast, with clumps of mangrove and forests of moss-dressed, strange trees appearing low in the distance. The steamer finds her way among the reedy islands like a duck, and I step on a rickety wharf. A few steps more take me to a rickety town, Fernandina. I discover a baker, buy some bread, and without asking a single question, make for the shady, gloomy groves. . . . Salt marshes, belonging more to the sea than to the land; with groves here and there, green and unflowered, sunk to the shoulders in sedges and rushes; with trees farther back, ill defined in their boundary, and instead of rising in hilly waves and swellings, stretching inland in low water-like levels. . . .

Florida is so watery and vine-tied that pathless wanderings are not easily possible in any direction. I started to cross the State by a gap hewn for the locomotive, walking sometimes between the rails, stepping from tie to tie, or walking on the strip of sand at the sides, gazing into the mysterious forest, Nature's own. It is impossible to write the dimmest picture of plant grandeur so redundant, unfathomable.

Short was the measure of my walk today. A new, canelike grass, or big lily, or gorgeous flower belonging to tree or vine, would catch my attention, and I would throw down my bag and press and splash through the coffee-brown water for specimens. Frequently I sank deeper and deeper until compelled to turn back and make the attempt in another and still another place. Oftentimes I was tangled in a labyrinth of armed vines like a fly in a spider-web. At all times, whether wading or climbing a tree for specimens of fruit, I was overwhelmed with the vastness and unapproachableness of the great guarded sea of sunny plants.

The eastern side of Cumberland Island National Seashore has the nicest dunes I know. But since the route of Muir's steamer was mostly west of the sea islands, down the Intracoastal Waterway, the passengers probably missed the dunes.

91

This is a close-up of the leaves of a cabbage palmetto (*Sabal palmetto*). A member of the palm family, the tree grows in a narrow strip along the coast from southeastern North Carolina to Georgia and in many places in Florida.

Magnolia grandiflora I had seen in Georgia; but its home, its better land, is here. Its large dark-green leaves, glossy bright above and rusty brown beneath, gleam and mirror the sunbeams most gloriously among countless flower-heaps of the climbing, smothering vines. It is bright also in fruit and more tropical in form and expression than the orange. It speaks itself a prince among its fellows.

Occasionally, I came to a little strip of open sand, planted with pine (*Pinus palustris* or *Cubensis*). Even these spots were mostly wet, though lighted with free sunshine, and adorned with purple liatris, and orange-colored *Osmunda cinnamomea*. But the grandest discovery of this great wild day was the palmetto.

I was meeting so many strange plants that I was much excited, making many stops to get specimens. But I could not force my way far through the swampy forest, although so tempting and full of promise. Regardless of water snakes or insects, I endeavored repeatedly to force a way through the tough vine-tangles, but seldom succeeded in getting farther than a few hundred yards.

It was while feeling sad to think that I was only walking on the edge of the vast wood, that I caught sight of the first palmetto in a grassy place, standing almost alone. A few magnolias were near it, and bald cypresses, but it was not shaded by them. They tell us that plants are perishable, soulless creatures, that only man is immortal, etc.; but this, I think, is something that we know very nearly nothing about. Anyhow, this palm was indescribably impressive and told me grander things than I ever got from human priest.

This vegetable has a plain gray shaft, round as a broom-handle, and a crown of varnished channeled leaves. It is a plainer plant than the humblest of Wisconsin oaks; but, whether rocking and rustling in the wind or poised thoughtful and calm in the sunshine, it has a power of expression not excelled by any plant high or low that I have met in my whole walk thus far.

This, my first specimen, was not very tall, only about twenty-five feet high, with fifteen or twenty leaves, arching equally and evenly all around. Each leaf was about ten feet in length, the blade four feet, the stalk six. The leaves are channeled like half-open clams and are highly polished, so that they reflect the sunlight like glass. The undeveloped leaves on the top stand erect, closely folded, all together forming an oval crown over which the

It was just by chance that while I was canoeing one day with photo-journalist Allan Horton he beached the canoe right by these bracket fungi (*Polyporus adustus*), a whole development of botanical "garden apartments."

Muir must have seen bracken ferns (*Pteridium aquilinum*) on his way across northern Florida because it is our country's commonest species. This example was growing on the banks of the Ichetucknee River.

Muir noted that he saw cinnamon ferns (*Osmunda cinnamomea*) in both Tennessee and Florida. Allan Horton spotted these plants while we were exploring.

tropic light is poured and reflected from its slanting mirrors in sparks and splinters and long-rayed stars.

I am now in the hot gardens of the sun, where the palm meets the pine, longed and prayed for and often visited in dreams, and, though lonely to-night amid this multitude of strangers, strange plants, strange winds blowing gently, whispering, cooing, in a language I never learned, and strange birds also, everything solid or spiritual full of influences that I never before felt, yet I thank the Lord with all my heart for his goodness in granting me admission to this magnificent realm.

October 16. Last evening when I was in the trackless woods, the great mysterious night becoming more mysterious in the thickening darkness, I gave up hope of finding food or a house bed, and searched only for a dry spot on which to sleep safely. . . .

When I came to an open place where pines grew, it was about ten o'clock, and I thought that now at last I would find dry ground. But even the sandy barren was wet, and I had to grope in the dark a long time, feeling the ground with my hands when my feet ceased to plash, before I at last discovered a little hillock dry enough to lie down on. I ate a piece of bread that I fortunately had in my bag, drank some of the brown water about my precious hillock, and lay down. The noisiest of the unseen witnesses around me were the owls, who pronounced their gloomy speeches with

93

profound emphasis, but did not prevent the coming of sleep to heal weariness.

In the morning I was cold and wet with dew, and I set out breakfastless. Flowers and beauty I had in abundance, but no bread. A serious matter is this bread which perishes, and, could it be dispensed with, I doubt if civilization would ever see me again. I walked briskly, watching for a house, as well as the grand assemblies of novel plants.

Near the middle of the forenoon I came to a shanty where a party of loggers were getting out long pines for ship spars. They were the wildest of all the white savages I have met. The long-haired ex-guerrillas of the mountains of Tennessee and North Carolina are uncivilized fellows; but for downright barbarism these Florida loggers excel. Nevertheless, they gave me a portion of their yellow pork and hominy without either apparent hospitality or a grudge, and I was glad to escape to the forest again. . . . Arrived at a place on the margin of a stagnant pool where an alligator had been rolling and sunning himself. "See," said a man who lived here, "see, what a track that is! He must have been a mighty big fellow. Alligators wallow like hogs and like to lie in the sun. I'd like a shot at that fellow." Here followed a long recital of bloody combats with the scaly enemy, in many of which he had, of course, taken an important part. . . .

I never in all my travels saw more than one, though they are said to be abundant in most of the swamps, and frequently attain a length of nine or ten feet. It is reported, also, that they are very savage, oftentimes attacking men in boats. These independent inhabitants of the sluggish waters of this low coast cannot be called the friends of man, though I heard of one big fellow that was caught young and was partially civilized and made to work in harness.

Many good people believe that alligators were created by the Devil, thus accounting for their all-consuming appetite and ugliness. But doubtless these creatures are happy and fill the place assigned them by the great Creator of us all. Fierce and cruel they appear to us, but beautiful in the eyes of God. They, also, are his children, for He hears their cries, cares for them tenderly, and provides their daily bread.

Perhaps the travelers were fortunate enough to experience a sunrise over a wild and deserted beach the morning of October 15th before they landed in Fernandina.

The antipathies existing in the Lord's animal family must be wisely planned, like balanced repulsion and attraction in the mineral king-

94

dom. How narrowly we selfish, conceited creatures are in our sympathies! how blind to the rights of all the rest of creation! With what dismal irreverence we speak of our fellow mortals! Though alligators, snakes, etc., naturally repel us, they are not mysterious evils. They dwell happily in these flowery wilds, are part of God's family, unfallen, undepraved, and cared for with the same species of tenderness and love as is bestowed on angels in heaven or saints on earth.

I think that most of the antipathies which haunt and terrify us are morbid productions of ignorance and weakness. I have better thoughts of those alligators now that I have seen them at home. . . .

Found a beautiful lycopodium to-day, and many grasses in the dry sunlit places called "barrens," "hummocks," "savannas," etc. Ferns also are abundant. What a flood of heat and light is daily poured out on these beautiful openings and intertangled woods! "The land of the sunny

By the side of Lake Orange. The flowers are southern blue-flag iris (*Iris virginica*). In this vicinity I found many vistas that would have pleased Muir and many, many others that would not.

South," we say, but no part of our diversified country is more shaded and covered from sunshine. Many a sunny sheet of plain and prairie break the continuity of the forests of the North and West, and the forests themselves are mostly lighted also, pierced with direct ray lances, or [the sunlight] passing to the earth and the lowly plants in filtered softness through translucent leaves. But in the dense Florida forests sunlight cannot enter. It falls on the evergreen roof and rebounds in long silvery lances and flashy spray. In many places there is not light sufficient to feed a single green leaf on these dark forest floors. All that the eye can reach is just a maze of tree stems and crooked leafless vine strings. All the flowers, all the verdure, all the glory is up in the light.

The streams of Florida are still young, and in many places are untraceable. I expected to find these streams a little discolored from the vegetable matter that I knew they must contain, and I was sure that in so flat a country I should not find any considerable falls or long rapids. The streams of upper Georgia are almost unapproachable in some places on account of luxuriant bordering vines, but the banks are nevertheless high and well defined. Florida streams are not yet possessed of banks and braes and definite channels. Their waters in deep places are black as ink, perfectly opaque, and glossy on the surface as if varnished. It often is difficult to ascertain which way they are flowing or creeping, so slowly and so widely do they circulate through the tree-tangles and swamps of the woods. The flowers here are strangers to me, but not more so than the rivers and lakes. Most streams appear to travel through a country with thoughts and plans for something beyond. But those of Florida are at home, do not appear to be traveling at all, and seem to know nothing of the sea.

October 17. Found a small, silvery-leafed magnolia, a bush ten feet high. Passed through a good many miles of open level pine barrens, as bounteously lighted as the "openings" of Wisconsin. The pines are rather small, are planted sparsely and pretty evenly on these sandy flats not long risen from the sea. Scarcely a specimen of any other tree is to be found associated with the pine. But there are some thickets of the little saw palmettos and a magnificent assemblage of tall grasses, their splendid panicles waving grandly in the warm wind, and making low tuneful changes in the glistening light that is flashed from their bent stems.

Not a pine, not a palm, in all this garden excels these stately grass plants in beauty of wind-waving gestures. Here are panicles that are one mass of refined purple; others that have flowers as yellow as ripe oranges, and stems polished and shining like steel wire. Some of the species are

Muir walked through level pine barrens as he worked his way west across Florida, and spotted saw-palmettos (*Serenoa repens*) growing among the pines. This scene is near Gainesville.

Nearing the Gulf of Mexico I discovered this clump of lady lupine (*Lupinus villosus*) growing in sandy soil. It was a new plant to me, and I learned it belongs to the bean family.

grouped in groves and thickets like trees, while others may be seen waving without any companions in sight. Some of them have wide-branching panicles like Kentucky oaks, others with a few tassels of spikelets dropping from a tall, leafless stem. But all of them are beautiful beyond the reach of language. I rejoice that God has "so clothed the grass of the field." How strangely we are blinded to beauty and color, form and motion, by comparative size! For example, we measure grasses by our own stature and by the height and bulkiness of trees. But what is the size of the greatest man, or the tallest tree that ever overtopped a grass! Compared with other things in God's creation the difference is nothing. We all are only microscopic animalcula.

October 18. Am walking on land that is almost dry. The dead levels are interrupted here and there by sandy waves a few feet in height. It is said that not a point in all Florida is more than three hundred feet above sea-level — a country where but little grading is required for roads, but much bridging, and boring of many tunnels through forests. . . .

A few miles farther on I came to a cottonfield, to patches of sugar cane carefully fenced, and some respectable-looking houses with gardens.

These little fenced fields look as if they were intended to be for plants what cages are for birds. Discovered a large treelike cactus in a dooryard; a small species was abundant on the sand-hillocks. Reached Gainesville late in the night. . . .

Gainesville is rather attractive — an oasis in the desert, compared with other villages. It gets its life from the few plantations located about it on dry ground that rises islandlike a few feet above the swamps. Obtained food and lodging at a sort of tavern.

October 19. Dry land nearly all day. Encountered limestone, flint, coral, shells, etc. Passed several thrifty cotton plantations with comfortable residences, contrasting sharply with the squalid hovels of my first days in Florida. Found a single specimen of a handsome little plant, which at once, in some mysterious way, brought to mind a young friend in Indiana. How wonderfully our thoughts and impressions are stored! There is that in the glance of a flower which may at times control the greatest of creation's braggart lords.

One of the rewards of wading in Florida swamps was photographing these trees in San Felasco Hammock—water ash (*Fraxinus caroliniana*) and water-elm (*Planera aquatica*). San Felasco Hammock has fortunately been purchased by the state of Florida for preservation, study, and research.

The magnolia is much more abundant here. It forms groves and almost exclusively forests the edges of ponds and the banks of streams. The easy, dignified simplicity of this noble tree, its plain leaf endowed with superb richness of color and form, its open branches festooned with graceful vines and tillandsia, its showy crimson fruit, and its magnificent fragrant white flowers make the magnolia the most lovable of Florida trees.

Discovered a great many beautiful polygonums, petalostemons, and yellow leguminous vines. Passed over fine sunny areas of the long-leafed and Cuban pines, which were everywhere accompanied by fine grasses and solidagoes. Wild orange groves are said to be rather common here, but I have seen only limes growing wild in the woods. . . .

October 20. Swamp very dense during this day's journey. Almost one continuous sheet of water covered with aquatic trees and vines. No stream that I crossed to-day appeared to have the least idea where it was going. Saw an alligator plash into the sedgy brown water by the roadside from an old log.

Arrived at night at the house of Captain Simmons, one of the very few scholarly, intelligent men that I have met in Florida. He had been an officer in the Confederate army in the war and was, of course, prejudiced against the North, but polite and kind to me, nevertheless. Our conversation, as we sat by the light of the fire, was on the one great question, slavery and its concomitants. I managed, however, to switch off to something more congenial occasionally — the birds of the neighborhood, the animals, the climate, and what spring, summer, and winter are like in these parts.

About the climate, I could not get much information, as he had always lived in the South and, of course, saw nothing extraordinary in weather to which he had always been accustomed. But in speaking of animals, he at once became enthusiastic and told many stories of hairbreadth escapes, in the woods about his house, from bears, hungry alligators, wounded deer, etc. "And now," said he, forgetting in his kindness that I was from the hated North, "you must stay with me a few days. Deer are abundant. I will lend you a rifle and we'll go hunting. I hunt whenever I wish venison, and I can get it about as easily from the woods near by as a shepherd can get mutton out of his flock. And perhaps we will see a bear, for they are far from scarce here, and there are some big gray wolves, too."

If Muir had landed on a coastal island, he might have found a swamp like this glorious one on Ossabaw, thoroughly draped with Spanish moss (*Tillandsia usneoides*).

I expressed a wish to see some large alligators. "Oh, well," said he, "I can take you where you will see plenty of those fellows, but they are not much to look at. I once got a good look at an alligator that was lying at the bottom of still, transparent water, and I think that his eyes were the most impressively cold and cruel of any animal I have seen. Many alligators go out to sea among the keys. These sea alligators are the largest and most ferocious, and sometimes attack people by trying to strike them with their tails when they are out fishing in boats.

"Another thing I wish you to see," he continued, "is a palmetto grove on a rich hummock a few miles from here. The grove is about seven miles in length by three in breadth. The ground is covered with long grass, uninterrupted with bushes or other trees. It is the finest grove of palmettos I have ever seen and I have oftentimes thought that it would make a fine subject for an artist."

I concluded to stop — more to see this wonderful palmetto hummock than to hunt. Besides, I was weary and the prospect of getting a little rest was a tempting consideration after so many restless nights and long, hard walks by day.

October 21. Having outlived the sanguinary hunters' tales of my loquacious host, and breakfasted sumptuously on fresh venison and "caller" fish from the sea, I set out for the grand palm grove. I had seen these dazzling sun-children in every day of my walk through Florida, but they were usually standing solitary, or in groups of three or four; but to-day I was to see them by the mile. The captain led me a short distance through his corn field and showed me a trail which would conduct me to the palmy hummock. He pointed out the general direction, which I noted upon my compass.

I was unable to learn the location of Captain Simmons's house, but I think it may have been near Archer. And for months I searched for a place that resembled the fine palmetto grove Simmons directed Muir to. Then by chance I came upon this rare, thick hammock of cabbage palmettos and grasses only a few miles from Gainesville. None of the local residents I'd talked to seemed aware of its existence.

"Now," said he, "at the other side of my farthest field you will come to a jungle of catbriers, but will be able to pass them if you manage to keep the trail. You will find that the way is not by any means well marked, for in passing through a broad swamp, the trail makes a good many abrupt turns to avoid deep water, fallen trees, or impenetrable thickets. You will have to wade a good deal, and in passing the water-covered places you will have to watch for the point where the trail comes out on the opposite side."

I made my way through the briers, which in strength and ferocity equaled those of Tennessee, followed the path through all of its dim waverings, waded the many opposing pools, and, emerging suddenly from the leafy darkness of the swamp forest, at last stood free and unshaded on the border of the sun-drenched palm garden. It was a level area of grasses and sedges, smooth as a prairie, well starred with flowers, and bounded like a clearing by a wall of vine-laden trees.

The palms had full possession and appeared to enjoy their sunny home. There was no jostling, no apparent effort to outgrow each other. Abundance of sunlight was there for every crown, and plenty to fall between. I walked enchanted in their midst. What a landscape! Only palms as far as the eye could reach! Smooth pillars rising from the grass, each capped with a sphere of leaves, shining in the sun as bright as a star. The silence and calm were as deep as ever I found in the dark, solemn pine woods of Canada, and that contentment which is an attribute of the best of God's plant people was as impressively felt in this alligator wilderness as in the homes of the happy, healthy people of the North.

The admirable Linnaeus calls palms "the princes of the vegetable world." I know that there is grandeur and nobility in their character, and that there are palms nobler far than these. But in rank they appear to me to stand below both the oak and the pine. The motions of the palms, their gestures, are not very graceful. They appear to best advantage when perfectly motionless in the noontide calm and intensity of light. But they rustle and rock in the evening wind. I have seen grasses waving with far more dignity. And when our northern pines are waving and bowing in sign of worship with the winter storm-winds, where is the prince of palms that could have the conscience to demand their homage!

Members of this palm congregation were of all sizes with respect to their stems; but their glorious crowns were all alike. In development there is only the terminal bud to consider. The young palm of this species emerges from the ground in full strength, one cluster of leaves arched every way, making a sphere about ten or twelve feet in diameter. The outside lower leaves gradually become yellow, wither, and break off, the petiole snapping squarely across, a few inches from the stem. New leaves develop with wonderful rapidity. They stand erect at first, but gradually arch outward as they expand their blades and lengthen their petioles.

New leaves arise constantly from the center of the grand bud, while old ones break away from the outside. The splendid crowns are thus kept

103

about the same size, perhaps a little larger than in youth while they are yet on the ground. As the development of the central axis goes on, the crown is gradually raised on a stem of about six to twelve inches in diameter. This stem is of equal thickness at the top and at the bottom and when young is roughened with the broken petioles. But these petiole-stumps fall off and disappear as they become old, and the trunk becomes smooth as if turned in a lathe.

After some hours in this charming forest I started on the return journey before night, on account of the difficulties of the swamp and the brier patch. On leaving the palmettos and entering the vine-tangled, half-submerged forest I sought long and carefully, but in vain, for the trail, for I had drifted about too incautiously in search of plants. But, recollecting the direction that I had followed in the morning, I took a compass bearing and started to penetrate the swamp in a direct line.

Of course I had a sore weary time, pushing through the tanglement of falling, standing, and half-fallen trees and bushes, to say nothing of knotted vines as remarkable for their efficient army of interlocking and lancing prickers as for their length and the number of their blossoms. But these were not my greatest obstacles, not yet the pools and lagoons full of dead leaves and alligators. It was the army of cat-briers that I most dreaded. I knew that I would have to find the narrow slit of a lane before dark or spend the night with mosquitoes and alligators, without food or fire. The entire distance was not great, but a traveler in open woods can form no idea of the crooked and strange difficulties of pathless locomotion in these thorny, watery Southern tangles, especially in pitch darkness. I struggled hard and kept my course, leaving the general direction only when drawn aside by a plant of extraordinary promise, that I wanted for a specimen, or when I had to make the half-circuit of a pile of trees, or of a deep lagoon or pond.

In wading I never attempted to keep my clothes dry, because the water was too deep, and the necessary care would consume too much time. Had the water that I was forced to wade been transparent it would have lost much of its difficulty. But as it was, I constantly expected to plant my feet on an alligator, and therefore proceeded with strained caution. The opacity of the water caused uneasiness also on account of my inability to determine its depth. In many places I was compelled to turn back, after wading forty or fifty yards, and to try again a score of times before I succeeded in getting across a single lagoon.

105

At length, after miles of wading and wallowing, I arrived at the grand cat-brier encampment which guarded the whole forest in solid phalanx, unmeasured miles up and down across my way. Alas! the trail by which I had crossed in the morning was not to be found, and night was near. In vain I scrambled back and forth in search of an opening. There was not even a strip of dry ground on which to rest. Everywhere the long briers arched over to the vines and bushes of the watery swamp, leaving no standing-ground between them. I began to think of building some sort of a scaffold in a tree to rest on through the night, but concluded to make one more desperate effort to find the narrow track.

After calm, concentrated recollection of my course, I made a long exploration toward the left down the brier line, and after scrambling a mile or so, perspiring and bleeding, I discovered the blessed trail and escaped to dry land and the light. Reached the captain at sundown. Dined on milk and johnny-cake and fresh venison. Was congratulated on my singular good fortune and woodcraft, and soon after supper was sleeping the deep sleep of the weary and the safe.

October 22. This morning I was easily prevailed upon by the captain and an ex-judge, who was rusticating here, to join in a deer hunt. Had a delightful ramble in the long grass and flowery barrens. Started one deer but did not draw a single shot. The captain, the judge, and myself stood at different stations where the deer was expected to pass, while a brother of the captain entered the woods to arouse the game from cover. The one deer that he started took a direction different from any which this particular old buck had ever been known to take in times past, and in so doing was cordially cursed as being the "d——dest deer that ever ran unshot." To me it appeared as "d——dest" work to slaughter God's cattle for sport. "They were made for us," say these self-approving preachers; "for our food, our recreation, or other uses not yet discovered." As truthfully we might say on behalf of a bear, when he deals successfully with an unfortunate hunter, "Men and other bipeds were made for bears, and thanks be to God for claws and teeth so long."

Let a Christian hunter go to the Lord's woods and kill his well-kept beasts, or wild Indians, and it is well; but let an enterprising specimen of these proper, predestined victims go to houses and fields and kill the most worthless person of the vertical godlike killers, — oh! that is horribly unorthodox, and on the part of the Indians atrocious murder! Well, I have precious little sympathy for the selfish propriety of civilized man, and if a war of races should occur between the wild beasts and Lord Man, I would be tempted to sympathize with the bears.

6
Cedar Keys

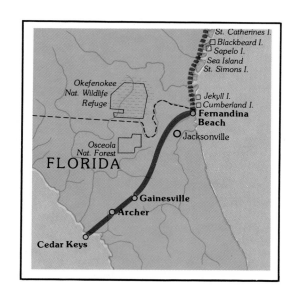

October 23. To-day I reached the sea. While I was yet many miles back in the palmy woods, I caught the scent of the salt sea breeze which, although I had so many years lived far from sea breezes, suddenly conjured up Dunbar, its rocky coast, winds and waves; and my whole childhood, that seemed to have utterly vanished in the New World, was now restored amid the Florida woods by that one breath from the sea. Forgotten were the palms and magnolias and the thousand flowers that enclosed me. I could see only dulse and tangle, long-winged gulls, the Bass Rock in the Firth of Forth, and the old castle, schools, churches, and long country rambles in search of birds' nests. I do not wonder that the weary camels coming from the scorching African deserts should be able to scent the Nile.

How imperishable are all the impressions that ever vibrate one's life! We cannot forget anything. Memories may escape the action of will, may sleep a long time, but when stirred by the right influence, though that influence be light as a shadow, they flash into full stature and life with everything in place. For nineteen years my vision was bounded by forests, but to-day, emerging from a multitude of tropical plants, I beheld the Gulf of Mexico stretching away unbounded, except by the sky. What dreams and speculative matter for thought arose as I stood on the strand, gazing out on the burnished, treeless plain!

But now at the seaside I was in difficulty. I had reached a point that I could not ford, and Cedar Keys had an empty harbor. Would I proceed down the peninsula to Tampa and Key West, where I would be sure to find

109

a vessel for Cuba, or would I wait here, like Crusoe, and pray for a ship. Full of these thoughts, I stepped into a little store which had a considerable trade in quinine and alligator and rattlesnake skins, and inquired about shipping, means of travel, etc.

The proprietor informed me that one of several sawmills near the village was running, and that a schooner chartered to carry a load of lumber to Galveston, Texas, was expected at the mills for a load. This mill was situated on a tongue of land a few miles along the coast from Cedar Keys, and I determined to see Mr. Hodgson, the owner, to find out particulars about the expected schooner, the time she would take to load, whether I would be likely to obtain passage on her, etc.

Found Mr. Hodgson at his mill. Stated my case and was kindly furnished the desired information. I determined to wait the two weeks likely to elapse before she sailed, and go on her to the flowery plains of Texas, from any of whose ports, I fancied, I could easily find passage to the West Indies. I agreed to work for Mr. Hodgson in the mill until I sailed, as I had but little money. He invited me to his spacious house, which occupied a shell hillock and commanded a fine view of the Gulf and many gems of palmy islets, called "keys," that fringe the shore like huge bouquets — not too big, however, for the spacious waters. Mr. Hodgson's family welcomed me with that open, unconstrained cordiality which is characteristic of the better class of Southern people.

At the sawmill a new cover had been put on the main driving pulley, which, made of rough plank, had to be turned off and smoothed. He asked me if I was able to do this job and I told him that I could. Fixing a rest and making a tool out of an old file, I directed the engineer to start the engine and run slow. After turning down the pulley and getting it true, I put a keen edge on a common carpenter's plane, quickly finished the job, and was assigned a bunk in one of the employees' lodging-houses.

The next day I felt a strange dullness and headache while I was botanizing along the coast. Thinking that a bath in the salt water might refresh me, I plunged in and swam a little distance, but this seemed only to make me feel worse. I felt anxious for something sour, and walked back to the village to buy lemons.

Thus and here my long walk was interrupted. I thought that a few days' sail would land me among the famous flower-beds of Texas. But the

Muir sailed in a skiff among the keys; I, too, sailed to some outlying keys, with Lee Belcher, who is in charge of the University of Florida's marine biology station at Sea Horse Key. One morning the waves were unusually high and before I knew it a camera in the bottom of the boat was actually floating. Then we landed at Sea Horse Key, and I forgot about my water-logged camera.

expected ship came and went while I was helpless with fever. The very day after reaching the sea I began to be weighed down by inexorable leaden numbness, which I resisted and tried to shake off for three days, by bathing in the Gulf, by dragging myself about among the palms, plants, and strange shells of the shore, and by doing a little mill work. I did not fear any serious illness, for I never was sick before, and was unwilling to pay attention to my feelings.

But yet heavier and more remorselessly pressed the growing fever, rapidly gaining on my strength. On the third day after my arrival I could not take any nourishment, but craved acid. Cedar Keys was only a mile or two distant, and I managed to walk there to buy lemons. On returning, about the middle of the afternoon, the fever broke on me like a storm, and before I had staggered halfway to the mill I fell down unconscious on the narrow trail among dwarf palmettos.

When I awoke from the hot fever sleep, the stars were shining, and I was at a loss to know which end of the trail to take, but fortunately, as it afterwards proved, I guessed right. Subsequently, as I fell again and again after walking only a hundred yards or so, I was careful to lie with my head in the direction in which I thought the mill was. I rose, staggered, and fell, I know not how many times, in delirious bewilderment, gasping and throbbing with only moments of consciousness. Thus passed the hours till after midnight, when I reached the mill lodging-house.

The watchman on his rounds found me lying on a heap of sawdust at the foot of the stairs. I asked him to assist me up the steps to bed, but he thought my difficulty was only intoxication and refused to help me. The mill hands, especially on Saturday nights, often returned from the village drunk. This was the cause of the watchman's refusal. Feeling that I must get to bed, I made out to reach it on hands and knees, tumbled in after a desperate struggle, and immediately became oblivious to everything.

I awoke at a strange hour on a strange day to hear Mr. Hodgson ask a watcher beside me whether I had yet spoken, and when he replied that I had not, he said: "Well, you must keep on pouring in quinine. That's all we

On Sea Horse Key I was able to photograph prickly-pear (*Opuntia humifusa*) and Spanish bayonet (*Yucca filamentosa*) growing side by side. Prickly-pear is the only cactus that is native east of the Mississippi.

can do." How long I lay unconscious I never found out, but it must have been many days. Some time or other I was moved on a horse from the mill quarters to Mr. Hodgson's house, where I was nursed about three months with unfailing kindness, and to the skill and care of Mr. and Mrs. Hodgson I doubtless owe my life. Through quinine and calomel — in sorry abundance — with other milder medicines, my malarial fever became typhoid. I had night sweats, and my legs became like posts of the temper and consistency of clay on account of dropsy. So on until January, a weary time.

As soon as I was able to get out of bed, I crept away to the edge of the wood, and sat day after day beneath a moss-draped live-oak, watching birds feeding on the shore when the tide was out. Later, as I gathered some strength, I sailed in a little skiff from one key to another. Nearly all the shrubs and trees here are evergreen, and a few of the smaller plants are in flower all winter. The principal trees on this Cedar Key are the juniper, long-leafed pine, and live-oak. All of the latter, living and dead, are heavily draped with tillandsia, like those of Bonaventure. The leaf is oval, about two inches long, three fourths of an inch wide, glossy and dark green above, pale beneath. The trunk is usually much divided, and is extremely unwedgeable. . . .

The live-oaks of these keys divide empire with the long-leafed pine and palmetto, but in many places on the mainland there are large tracts exclusively occupied by them. Like the Bonaventure oaks they have the upper side of their main spreading branches thickly planted with ferns, grasses, small saw palmettos, etc. There is also a dwarf oak here, which forms dense thickets. The oaks of this key are not, like those of the Wisconsin openings, growing on grassy slopes, but stand, sunk to the shoulders, in flowering magnolias, heathworts, etc.

During my long sojourn here as a convalescent I used to lie on my back for whole days beneath the ample arms of these great trees, listening to the winds and the birds. There is an extensive shallow on the coast, close by, which the receding tide exposes daily. This is the feeding-ground of thousands of waders of all sizes, plumage, and language, and they make a lively picture and noise when they gather at the great family board to eat their daily bread, so bountifully provided for them.

Their leisure in time of high tide they spend in various ways and places. Some go in large flocks to reedy margins about the islands and wade and stand about quarrelling or making sport, occasionally finding a stray mouthful to eat. Some stand on the mangroves of the solitary shore, now and then plunging into the water after a fish. Some go long journeys inland, up creeks and inlets. A few lonely old herons of solemn look and wing retire to favorite oaks. It was my delight to watch those old white sages of immaculate feather as they stood erect drowsing away the dull hours between tides, curtained by long skeins of tillandsia. White-bearded hermits gazing dreamily from dark caves could not appear more solemn or more becomingly shrouded from the rest of their fellow beings.

One of the characteristic plants of these keys is the Spanish bayonet, a species of yucca, about eight or ten feet in height, and with a trunk three or four inches in diameter when full grown. It belongs to the lily family and develops palmlike from terminal buds. The stout leaves are very rigid, sharp-pointed and bayonet-like. By one of these leaves a man might be as seriously stabbed as by an army bayonet, and woe to the luckless wanderer who dares to urge his way through these armed gardens after dark. Vegetable cats of many species will rob him of his clothes and claw his flesh, while dwarf palmettos will saw his bones, and the bayonets will glide to his joints and marrow without the smallest consideration for Lord Man. . . .

Cedar Key is two and one half or three miles in diameter and its highest point is forty-four feet above mean tide-water. It is surrounded by scores of other keys, many of them looking like a clump of palms, arranged like a tasteful bouquet, and placed in the sea to be kept fresh. Others have quite a sprinkling of oaks and junipers, beautifully united with vines. Still others consist of shells, with a few grasses and mangroves, circled with a rim of rushes. Those which have sedgy margins furnish a favorite retreat for countless waders and divers, especially for the pelicans that frequently whiten the shore like a ring of foam.

It is delightful to observe the assembling of these feathered people from the woods and reedy isles; herons white as wave-tops, or blue as the sky, winnowing the warm air on wide quiet wing; pelicans coming with baskets to fill, and the multitude of smaller sailors of the air, swift as swallows, gracefully taking their places at Nature's family table for their daily bread. Happy birds!

The mockingbird is graceful in form and a fine singer, plainly dressed, rather familiar in habits, frequently coming like robins to doorsills for crumbs — a noble fellow, beloved by everybody. Wild geese are abundant in winter, associated with brant, some species of which I have never seen in the North. Also great flocks of robins, mourning doves, bluebirds, and the

delightful brown thrashers. A large number of the smaller birds are fine singers. Crows, too, are here, some of them cawing with a foreign accent. The common bob-white quail I observed as far south as middle Georgia.

Lime Key . . . is a fair specimen of the Florida keys on this part of the coast. The cactus, *Opuntia* . . . is from the above-named key, and is abundant there. The fruit, an inch in length, is gathered, and made into a sauce, of which some people are fond. This species forms thorny, impenetrable thickets. One joint that I measured was fifteen inches long.

The mainland of Florida is less salubrious than the islands, but no portion of this coast, nor of the flat border which sweeps from Maryland to Texas, is quite free from malaria. All the inhabitants of this region, whether black or white, are liable to be prostrated by the ever-present fever and ague, to say nothing of the plagues of cholera and yellow fever that come and go suddenly like storms, prostrating the population and cutting gaps in it like hurricanes in woods.

The world, we are told, was made especially for man — a presumption not supported by all the facts. A numerous class of men are painfully astonished whenever they find anything, living or dead, in all God's universe, which they cannot eat or render in some way what they call useful to themselves. They have precise dogmatic insight of the intentions of the Creator, and it is hardly possible to be guilty of irreverence in speaking of *their* God any more than of heathen idols. He is regarded as a civilized, law-abiding gentleman in favor either of a republican form of government or of a limited monarchy; believes in the literature and language of England; is a warm supporter of the English constitution and Sunday schools and missionary societies; and is as purely a manufactured article as any puppet of a half-penny theater.

White ibis (*Eudocimus albus*) aloft. Muir watched many species of shorebirds at Cedar Key as he convalesced after his illness.

With such views of the Creator it is, of course, not surprising that erroneous views should be entertained of the creation. To such properly trimmed people, the sheep, for example, is an easy problem — food and clothing "for us," eating grass and daisies white by divine appointment for this predestined purpose, on perceiving the demand for wool that would be occasioned by the eating of the apple in the Garden of Eden.

In the same pleasant plan, whales are storehouses of oil for us, to help out the stars in lighting our dark ways until the discovery of the Pennsylvania oil wells. Among plants, hemp, to say nothing of the cereals, is a case of evident destination for ships' rigging, wrapping packages, and hanging the wicked. Cotton is another plain case of clothing. Iron was made for hammers and ploughs, and lead for bullets; all intended for us. And so of other small handfuls of insignificant things.

But if we should ask these profound expositors of God's intentions, how about those man-eating animals — lions, tigers, alligators — which smack their lips over raw man? Doubtless man was intended for food and drink for all these? Oh, no! Not at all! . . .

Now, it never seems to occur to these far-seeing teachers that Nature's object in making animals and plants might possibly be first of all the happiness of each one of them, not the creation of all for the happiness of one. Why should man value himself as more than a small part of the one great unit of creation? And what creature of all that the Lord has taken the pains to make is not essential to the completeness of that unit — the cosmos? The universe would be incomplete without man; but it would also be incomplete without the smallest transmicroscopic creature that dwells beyond our conceitful eyes and knowledge.

From the dust of the earth, from the common elementary fund, the Creator has made *Homo sapiens*. From the same material he has made *every* other creature, however noxious and insignificant to us. They are earth-born companions and our fellow mortals. The fearfully good, the orthodox, of this laborious patchwork of modern civilization cry "Heresy" on every one whose sympathies reach a single hair's breadth beyond the boundary epidermis of our own species. Not content with taking all of earth, they also claim the celestial country as the only ones who possess the kind of souls for which that imponderable empire was planned.

This star, our own good earth, made many a successful journey around the heavens ere man was made, and whole kingdoms of creatures enjoyed existence and returned to dust ere man appeared to claim them. After human beings have also played their part in Creation's plan, they too may disappear without any general burning or extraordinary commotion whatever.

Plants are credited with but dim and uncertain sensation, and minerals with positively none at all. But why may not even a mineral arrangement of matter be endowed with sensation of a kind that we in our blind exclusive perfection can have no manner of communication with?

But I have wandered from my object. I stated a page or two back that man claimed the earth was made for him, and I was going to say that venomous beasts, thorny plants, and deadly diseases of certain parts of the earth prove that the whole world was not made for him. When an animal from a tropical climate is taken to high latitudes, it may perish of cold, and we say that such an animal was never intended for so severe a climate. But when man betakes himself to sickly parts of the tropics and perishes, he cannot see that he was never intended for such deadly climates. No, he will rather accuse the first mother of the cause of the difficulty, though she may never have seen a fever district; or will consider it a providential chastisement for some self-invented form of sin.

Furthermore, all uneatable and uncivilizable animals, and all plants which carry prickles, are deplorable evils which, according to closet researches of clergy, require the cleansing chemistry of universal planetary combustion. But more than aught else mankind requires burning, as being in great part wicked, and if that transmundane furnace can be so applied and regulated as to smelt and purify us into conformity with the rest of the terrestrial creation, then the tophetization of the erratic genus Homo were a consummation devoutly to be prayed for. But, glad to leave these ecclesiastical fires and blunders, I joyfully return to the immortal truth and immortal beauty of Nature.

Watching my last sunrise at Cedar Key, I became convinced that had Muir not caught malaria and so been unable to book passage to South America, he would indeed have survived the daring float trip he'd planned down the Amazon River, no matter what hazards and hardships he encountered along the way.

Acknowledgments

I am most grateful to the following individuals for their kindness, information, and assistance while I worked on this book. If there are some people whom I have overlooked, they are no less included in my appreciation.

Eleanor, Clifford, and Justin West   Anne and Charles Wood

Ethel and Lester Morgan   Willard Trask   Clermont Lee

Kathryn and Allison Wade   Frances and Harry Purvis

Larry Williams   Mr. and Mrs. L.M. Hutchinson

Jan and Edgar Monroe   Carroll Scruggs   Peggy and Allan Worms

John Warren   Jim Jackson   Christine Williams

Joyce and Don Murray   Helen and Ian Hood

Martha and Allan Horton   Esta and Lee Belcher   Marie Mellinger

Anna and Coy Hanson, National Park Service

Frank Mayfield, U.S. Forest Service   Don Fig, U.S. Forest Service

Zeb Palmer, U.S. Forest Service   C.W. Smith, U.S. Forest Service

Harvey F. Price, U.S. Forest Service

I only went out for a walk
and finally concluded to stay out till sundown,
for going out, I found, was really going in.

John Muir